# Citrus Classification

R.C. Woodford

BIOTECH
2016
Biotech Books
Delhi - 110 035

ISBN 81-7622-108-2

*Published by* : **Biotech Books**
1123/74, Tri Nagar,
Delhi-110 035
Ph.: 27382765
e-mail : biotechbooks@yahoo.co.in

*Showroom* : 4762-63/23, Ansari Road
New Delhi-110 002
Ph.: 23245578, 23244987

*Laser Composed by* : **Vaishnav Graphics & Systems**
New Delhi-110 055

*Printed at* : **Chawla Offset Printers**
Delhi-110 052

**PRINTED IN INDIA**

# Foreword

This Monograph on Citrus Classification has behind it well over ten years of arduous and patient labour by the authors. The work was sponsored and started by my predecessor, Dr. S.K. Mitra. It gives me great pleasure to witness its completion just before I retire.

Assam is so very rich in citrus wealth that the correct identification of the numerous forms, and the assessment of their horticultural potential, is necessary in order to utilize these materials for improvement of the citrus industry.

The present work is not based on an itinerant study, but is built up on the very intimate association of the authors with the numerous plant materials collected from all over the province and studied under both field and laboratory conditions. I am glad to notice that the writers have incorporated in the text useful horticultural information as well as botanical details.

The work will, I hope, lead to further investigation and research from horticultural, cytological, bio-chemical and other connected aspects.

**R.C. Woodford**

*Director of Agriculture, Assam*

# CONTENTS

# Chapter 1

# INTRODUCTION

Assam has been growing numerous forms of Citrus from a hoary past and it also abounds in wild forms of this genus. Hence it is considered by many to be the 'creater centre' of citrus flora of India. No other province in India probably has a richer natural array of diverse forms of Citrus fruits growing with so much ease in the hills and dales without very much human aid. It may be that the diversity of her soils ranging from the alluvials of the plains to the lateritic formations of the hills, the rainfall varying from 40 to 500 in., the altitude varying from 100 to 8,000 ft., and varying humidity and forcing temperature obtaining in different parts of the province offer optimum condition of growth for different Citrus varieties to flourish and multiply in one part of the province or the other, making it a natural repository for this important group of fruits. In a naturally crosspollinated genus like the Citrus, Nature is eventually active in her dynamic efforts to create ever more new forms and Assam with its favourable conditions of soil and climate possibly offer unique opportunities for the perpetuation of many of these.

The wide adaptability of Citrus fruits of Assam is reflected in their general distribution in three well-defined topographical situations, namely the Brahmaputra and Surma Valley plains between 200 to 500 ft., sub-montane regions betwen 500 to 2,000 ft., and high hills above 2,000 ft. The mandarin *Khasi* oranges *(Citrus reticulata) par excellence* of commerce, are mostly confined to the sub-montane regions though they are sometimes grown in higher elevations even up to 5,000 ft., with a marked deterioration in size and quality of the fruits. The limes, the lemons, the citrons, and the pummelos predominate in the plains and to smaller extent in the sub-montane tracts. There are some species like *Citrus ichangensis* and *Citrus latipes* which more or less grow exclusively in high altitudes over 4,000 ft.

Where the diversity of forms is great the room for confusion in regard to identity is also equally great and no other important genus has probably been more liable to nomenclatural confusion than the Citrus which yet continues to be a riddle awaiting an acceptable solution in the hands of the systematists. Even in India the same fruit parades under different names in different provinces and except probably the *santara* which is a well recognized name in trade for mandarin orange, no other variety passes under a common name in the whole of India. In Assam with its babel of tongues in vogue in different parts of the province, particularly in the hill tracts, the situation is still worse. Even within the same district a variety may often be known by different names as will be seen in the chapters dealing with taxonomical description of the different fruits. The necessity of classifying this motley crowd of the Citrus varieties, long in existence in Assam, into a well recognised systematic pattern for easy and proper identification, has long been recognized by the Assam Agricultural Department and it was with the inauguration of the Citrus Fruit Research Scheme at Buraihat, financed jointly by the Indian Council of Agricultural Research and the Government of Assam, in 1938, that the work was taken up as one of the main items of the research programme. This has been carried on now for little over ten years. During this period considerable data have been collected and compiled and the present monograph aims at presenting the same to the readers.

The first pre-requisite for any work on classification is the collection of representative samples on which to base the study, for no systematic work can be deemed as complete unless it includes in its purview the the study of a large number of specimens collected from a wide range of soil and climatic conditions.

The work of collection was undertaken in Assam from 1938 and was carried on without respite for little over ten years, covering all important Citrus growing areas of the province, both in the hills and in the plains, with particular attention to the sub-montane tracts where most of the commercial Citrus plantations occur. The collection of wild specimens from the forests uninhabited by human being, was done mostly with the help of the officers of the Forest and the Political Departments of the province. The work was executed in a very thorough manner and there is hardly any important Citrus tract in the province situated between 100 ft. and 8,000 ft. which was not visited in quest of the Citrus varieties.

The collection of different specimens consisted of gathering both mature fruits and inflorescence with leaves. Wherever available both seedling and clonal progenies of different varieties were also collected. In some cases where no mature fruits or plants were obtainable, bud materials were procured and later on propagated on root stocks. For each speciman a minimum of two fruits was preserved in formalin and two herbarium sheets embodying inflorescence and leaves were also preserved in the laboratory. For detailed observational study a Culture Block was maintained representing at least four seedlings or clonal progenies of each specimen. During collection no attempt was made to eliminate the replicas unless the similarity was very obvious. This cautious handling of the materials, even at the risk of enormous multiplication of work, was adopted to minimise the chance of any new type or form slipping out. The enormity of the work may be judged by the fact that there are 1076 plants growing in the Culture Block representing 270 different samples.

Most of the systematic works done on Citrus in the past by different workers have been based on the study of dry herbarium specimens with its obvious limitations. This method of approach per force confines the study to limited morphological details of leaves, flowers and shrivelled fruits as mounted on herbarium sheets. Important as these are, they are by no means the only diagnostic features for arriving at reliable conclusions. The habit of growth, the mode of flowering and fruiting, etc., might sometimes differentiate two forms which are otherwise alike. The maintenance of a 'live museum of plants' in the field side by side with the herbarium specimens preserved in the laboratory as was done by us, gave opportunities for studying in minutest details both botanical and horticultural peculiarities of individual specimens. All the representative plants growing in the Culture Block have been studied with reference to numerous important taxonomic features, comprising of characteristics of growth, mode of flowering and fruiting, external characters of leaves and flowers and both external and internal characters of fruits. While recording the diagnostic details of leaves, flowers and fruits, only fresh samples representing a fair average of the variety under study have been considered, avoiding variable extremes. This was necessary as the variation in size of leaves, flowers and fruits in Citrus is quite considerable. This is particularly so in the case of those varieties which flower and fruit more than once in a year. The difference in size between the big summer leaves and the small winter ones, between the flowers and fruits of the normal season and those of the 'off season' is amazingly great. Even amongst the fruits of the same tree of the same season the extent of variation in size and juice content is observed to be very marked.

Apart from recording the taxonomical details, representative fruit samples of most of the varieties were also chemically analyzed for their sugar and citric acid content and data collected have been presented in an appendix so as to exhibit their difference in important chemical potentials.

All the taxonomic details and other informations have been systematically recorded in a 'descriptive blank' prepared on the model of H.J. Webber (*The Citrus Industry*. Vol. 1, Pp. 481-84) with some addition and alteration With regard to the description of the shape of the fruits and leaves, Hedrick (1925) and McCann (1947) respectively have generally been followed. This has ensured a more or less uniform and standardized descriptions and facilitated an easy comparison of the characters of different species and varieties under study. While describing the varieties under a given species in this treatise, the morphological characters of each have been stated fully, even at the cost of repetition so as to bring out the varietal peculiarities very prominently and also for easy comprehension. Though this will entail adding to the text of taxonomical descriptions, it would nonetheless, it is hoped, help in portraying a complete and self-contained picture of the variety under study.

No work on classification may be deemed as complete unless the important specific characters are supported by good coloured illustrations and photographs of the same. The assemblage of botanical terminologies, however precise they may be in their meaning, cannot by itself be expected to clear up all the confusions especially in a genus life Citrus with numerous 'off types'and intermediate forms. Hume (1941) rightly says 'Illustrations are an important part of any horticultural work–a single picture may tell more than a page or indeed several pages of prints'. This explains the reason for presenting the numerous coloured illustrations and photographs in support of the descriptive details of the different species and varieties. All the pictorial representations have been drawn to the scale as shown in the text.

The identification of different varieties and species was done by critically studying almost all available literature on the subject and comparing notes. Help was also taken of the Royal Botanic Gardens, Sibpur, Calcutta and the Forest Herbarium, Shillong for comparing our herbarium sheets, coloured illustrations and descriptive blanks with details duly filled in, with the identified type specimens as preserved in those herbaria. This, however, did not carry us far in the matter of the identification as the collections in both the herbaria are confined to a few well recognized species only.

## Discussion

Considerable work has been done on the classification of Citrus by various taxonomists in different parts of the world, notably in, the United States of America and Japan and a brief review of the same with reference to Indian Citrus and Assam Citrus in particular, is probably necessary to explain the basis of the nomenclature as adopted in this treatise. Ours is obviously a continuation of the kindred work done in the past. We can hardly claim any originality for the pattern of classification as adopted here, which unquestionably belongs to Swingle and Webber. We have, however, attempted to broadbase the pattern by recognizing some species not accepted by Swingle (1946) so as to fit in the Assam Citrus in a more easily recognizable and a natural way. A brief resume on the taxonomical works on Citrus as done by numerous authours is furnished below.

Roxburgh (1832) in the *Flora Indica* recognized five *Citrus species,* namely : (i) *C.acida,* (ii) *C.medica* (citron), (iii) *C.aurantium* (orange), (iv) *C.decumana* (shaddock), and (v) *C.inermis* (kumquat). He took no cognizance of the differences between limes and lemons and included both under the common species *C.acida.* To the most important commercial orange, *kamala* or *santra,* he gave the name *C.aurantium* which is now the accepted name for sour or bitter orange of which he makes no mention at all. He recognized the existence of shaddock in India and believed it to be of Batavian origin. He also listed kumquat like *Citrus* under a specific name *C.inermis* which was supposed to be imported from China. This is now recognized to belong to a separate genus, *Fortunella.*

Rafinesque (1838) in his book *Sylva Telluriana*, published in 1838, mentioned *C.combara* (combara), *C.costata (kamala or kolombok)*, *C.gongra (Gongra* of Bengal), *C. granulata* (Granular lemon of India), *C. karna* (*karna*), etc. but none of the later authors on Indian citrus flora had made any reference to his species. *C.combara* of Rafinesque is probably represented by *satkara* or *soh-kwit* of Assam, although Tanaka (1937, a) considers it to be similar to *C.macroptera* var. *combara* or *annamensis* Tan. The species *C.karna* is common in India and is represented by *soh-sarkar* of Assam and *kharna khatta* of the Punjab. This is the only species of Rafinesque which is referred to in occasional publications in India. *C.karna* was renamed and *C.dimorphocarpa* by Lushington(1910).

Hooker (1872) in his *Flora of British India* recognized only four species of *Citrus* to accommodate all the Citrus varieties of India. They are : (i) *C.medica,* (ii) *C.aurantium,* (iii) *C.hystrix,* and (iv) *C.decumana.* He did not accept the specific name *C.acida* of Roxburgh and instead he included citron, lemon and limes both acid and sweet forms, within the species *C.medica.* But he exclusively retained *acida* as varietal name for the sour lime of India. To Roxburgh's species of *Citrus* Hooker added *C. hystrix* and eliminated *C. inermis.* With regard to other speices such as *C.aurantium* and *C.decumana* he particularly followed Roxburgh. He placed sweet orange, mandarin oranges of the Khasi Hills, Garhwal, and Sikkim and also bitter or sour orange within a single species, *C.aurantium.* Here it might be pointed out that he recognized the bitter orange, and Bargamotte orange which Roxburgh did not mention in his Flora of India. About *C.decumana* (shaddock) Hooker reported that it was cultivated very commonly in India indicating that during the period of about 40 to 50 years after Roxburgh's account, this variety had fairly spread out all over India. Hooker's grouping of distinct botanical units into a few broad-compassed species like his predecessor is rather unsatisfactory to accommodate the great diversity of Indian materials. Kanjilal, Das, and Purkayastha (1936) followed Hooker in their classification of citrus in the *Flora of Assam.*

Brandis (1874) in *Forest Flora of North West and Central India* listed only three *Citrus* species namely (i) *C.medica,* (ii) *C.aurantium,* and (iii) *C.decumana.* Brandis again in his *Indian Trees* written in 1906, maintained the same number of species of Indian Citrus disregarding the botanical and horticultural peculiarities of limes, lemons and citrons which by no means can be grouped together into one specific rank, *C.medica*. Like his predecessors, he included loose-skinned orange, tight-skinned orange and bitter orange in a single species, *C.aurantium*. His system of Citrus classification cannot be considered an improvement over that of his earlier authors and is in fact a step retrograde. The system of Citrus nomenclature as adopted by Brandis has been basically adhered to by Duthie (1903), Cooke (1903), Talbot (1909), Parker (1918) in their works on genus *Citrus*.

Kurz (1877) in *Forest Flora of British Burma* enumerated only five species such as : (i) *C.medica*, (ii) *C.decumana*, (iii) *C.hystrix*, (iv) *C.aurantium*, and (v) *C.nobilis*. It may be pointed out here that Kurz definitely upheld the status of loose-skinned orange of commerce having assigned to it a specific name, *C.nobilis*. His botanicol descriptions of *C.nobilis* and *C.aurantium*, however, are inadequate to form a very clear idea of these species.

Bonavia (1888-90) was perhaps the first to study critically almost all the varieties of Citrus fruits of India and Ceylon. His study, though strictly speaking does not come within the purview of taxonomic classification, yet it attempts at placing the different varieties to well defined groups to which they naturally belong. This undoubtedly gives a clear idea of the varieties of India and Ceylon for future studies. His theory of 'elephantism' of forms particularly of the pummelo is fantastic and that of morphology of Citrus is highly imaginative and in most cases untenable in the light of the modern knowledge on the subject. Bonavia perhaps was the first in India to differentiate between the loose-skinned mandarin oranges and the tight-skinned *maltas*. He gave an independent status to the *santra* (Mandarin) orange of India and dismissed the idea that it was of Portuguese origin. He also initiated a discussion on the position of the intermediate forms particularly of lemons and citrons and his placing of such intermediate forms is an improvement upon the older conception of grouping. He, however, did not attempt to give any specific or varietal name to the cultivated varieties of Citrus of India or Ceylon except quoting in some cases obsolete scientific names, wherever possible, for identification of materials. His discussion on the origin of Indian Citrus varieties is very much informative and learned.

Watt (1889) in the *Dictionary of Economic Products of India* has compiled and elaborately discussed the works of earlier authors on citrus classification and has mentioned four species namely, *C.aurantium, C.decumana, C.medica* and *C.nobilis* to cover all the Indian varieties. The species *C.aurantium* has been divided into three varieties : (i) *aurantium* proper (sweet orange or China orange), (ii) *aurantium* var. *Bigardia,* and (iii) var. *Bergamia.* The *aurantium* proper has been further subdivided into four races : (a) *Santra,* (b) *Narangi or Keonla,* (c) *Malta* or Blood Red, and (d) Mandarin comprising of Sylhet or Khasi Hill orange and orange of Nagpur, apparently placing both tight-skinned sweet orange and loose skinned mandrian orange under one and the same group and making a distinction between *santra* (loose-skin) and mandarin (thin and smooth skin). From the list of vernacular names furnished by the author under *Aurantium* proper it appears that *rabab-tenga (C.grandis)* and *jora-tenga (C.medica)* of Assam, which are basically different from each other as well as from the mandarin oranges are classed together under the mandarin race. This shows that he was misinformed about the local names of the mandarin oranges of Assam.

Lushington (1910) in an article entitled 'The genus *Citrus*', published in the *Indian Forester,* 1910 recorded taxonomic details of nearly all the varieties of citrus which had been collected and studied by Bonavia in 1890. In that paper, while giving botanical names for various Citrus of India and Ceylon as studied by Bonavia, he recognized as many as twenty one species of Citrus which might be considered as a distinct departure from the earlier writers on Indian Citrus. His long list of species, with the exception of a few well established ones, had not been ultimately accepted by Western taxonomists. Tanaka (1937, a) is the only exception, who appears to have recognized some four species of Lushington such as *C.chrysocarpa* for santra orange, *C.jambhiri* for *jambhiri* or so called rough lemon, *C.megaloxycarpa* for *amilbed* or so called sour pummelo and *C.crenatifolia* for *keonla* or *kokni (C.mitis.)* With the exception of a few, most of the Citrus fruits detailed by Lushington may be absorbed in widely known species or varieties as the case may be. Lushington's classification, on the whole, is very much broad-based and some of his new species seem to deserve recognition.

Coit (1922) in his *Citrus Fruits* recognized ten species of *Citrus* and included *Trifoliata* and *Japonica* which are now considered to belong to different genera.

Osmasion (1927) in the *Forest Flora for Kumaon* mentions only about *C.medica,* meaning obviously citron. Although Kumaon is very rich in Citrus it is strange that he had not come across any other forms.

Cheema and Bhat (1934) in their *A Study of the Citrus Fruits of Bombay Presidency* lucidly discussed the present status of Citrus fruits of Bombay and sought to fit in the Bombay varieties with the different patterns of Citrus classification as advocated by Bonavia (1890), Coit (1922), Tanaka (1927) and Swingle (1919). This method of treatment has brought out the taxonomic position of the Bombay varieties of Citrus and cleared up much confusion as to the placing of the varietes to their proper belonging. It is not known why they have not discussed in ther interesting paper anything about the system of classification of the Citrus of Bombay as has been done by Dalzell and Gibson (1861) and also by Cooke (1903).

Tanaka (1933 and 1937) in *Proc. Jap. Assoc. Advan. Sci.* 10,576 accepted 15 'primary elements or species' for the genera *Poncirus, Fortunella* and *Citrus* of which 13 species constitute the genus *Citrus.* Later in 1937 in his *further Revision of Rutaceae–Aurantioideae of India and Ceylon'*, he enumerated as many as twenty-two different new species over his 'primary elements'. In support of his new species, no cogent reasons have been advanced by many of his new species, for example *C.limettioides, C.erythrosa, C.regulosa, V.reshmi,* etc., can conveniently be eliminated and absorbed in the older and well recognized species without practically changing their taxonomic status. His Citrus classification seems to be much more broad-based and liberal than is needed. With regard to the Indian Citrus he gave either a specific or a varietal rank to almost all the Citrus fruits he happened to come across apparently without going much into details. His recognition of innumerabe species and varieties of India and Ceylon, however, just goes to illustrate the richness of Citrus flora of this sub-continent.

Hume (1941) recognized only ten species of *Citrus* in his book *The Cultivation of Citrus Fruits.* Of the ten species, *C.maxima* (pummelo) and *C.nobilis* (mandarin orange) have been changed to *C.grandis* and *C.reticulata* respectively by Swingle (1946) in his recent publication. Hume's *C.mitis* Osbeck has been considered as a variety under the species *C.reticulata* by Webber (1946). Detailed diagnostic characters of these ten species have been given by him for proper understanding.

Swingle (1912 to 1946) published a number of taxonomic papers on tribes, sub-tribes, genera and species of the orange sub-family. The latest publication on the subject is perhaps the book *The Citrus Industry,* Vol.I, wherein about 345 pages have been devoted by him in a chapter captioned 'Botany of Citrus'. He deals elaborately on the orange sub-family. The scheme of his Citrus classification though to some extent conservative yet is comprehensive and based on personal experience of vast materials of both herbaria and living plants. His work undoubtedly gives a cear perspective of the specific and varietal conception within the limit of his accepted species. He recognized only sixteen species and eight botanical varieties of Citrus. The notable change of the botanical nomenclature as suggested and accepted by him is *C. reticulata* for mandarin orange, usually designated as *C.nobilis C.chrysocarpa, C.dcliciosa,* etc. The retention of the name *C.reticulata* as applied by Blanco in 1837, has been found to be justifiable by subsequent studies of Blanco's plants by Merrill (1918). Swingle also in 1914 in his '*Plantae Wilsomanae,*Pp. 141-149' clearly points out that true mandarin orange is different from *C.nobilis,* the 'King orange'.This has really removed the long drawn confusion of the correct nomenclature of the mandarin orange. He also favours retaining the name *C.limon* Burm., for true sour lemon of commerce instead of calling lemon by the name *C.limonia* Osbeck; because Osbeck, while building up this species had not true lemon in view. Similarly Swingle (1946) considers *C.grandis* (Linn.) Osbeck as the correct name for pummelo which is also called as *C.maxima* and *C.decumana.* Swingle also emphasises by quoting others that chemical composition, particularly the presence of different glucosides in the different species of *Citrus* (7 species have been mentioned) has definite significance in segregating one species from the other. This is indeed a new approach as it deals with the classification both from taxonomic and chemical points of view.

Webber (1946) in the *Citrus Industry,* Vol. I, has also devoted about 300 pages on 'cultivated varieties of Citrus' dealing with the general aspects of varietal classification. He has described taxonomically all the varieties of Citrus grown in the United States of America. The mode of his treatment of the cultivated horticultural varieties is clear and very much instructive. He has generally followed Swingle in his scheme of classification.

## Conclusion

From the above symposium it would be abundantly clear that most of the authors of Indian Citrus have confined themselves to a few broad-compassed (4 or 5) species totally disregarding the richness of Indian materials. Lushington (1910) and Tanaka (1937a) are the two exceptions who endeavoured to liberalise the number of species so as to fit n with the enormity of Indian forms, though undoubtedly some of their species could have been profitably reduced without altering basically their scheme of classification. The scheme of Swingle (1946) though it looks somewhat conservative in out-look, seems to be sufficiently comprehensive and elastic for general adoption as with certain minor adjustments it can be made to accommodate most of the known Indian varieties.

Generally, Swingle's system of classification has been adopted by us and accordingly the genus *Citrus* has been grouped into two sub-genera, *viz.*, (i) *Eucitrus* (cultivated edible Citrus) and (ii) *Papeda* (inedible Citrus). The latter has again been sub-divided in to two sections (i) *Eupapeda and* (ii) *Papedo-Citrus.* The different species have been arranged under the above groupings. Though we have adopted Swingle's Scheme in a general way yet in a few cases we have been obliged out of necessity to accept certain species not recognized by Swinlge, namely (i) *C.nobilis*, Lour, (ii) *C.karna* Raf., (iii) *C.limetta* (Risso) Lush., (iv) *C.jambhiri* Lush., and (v) *C.megaloxycarpa* Lush. With regard to these species referred to above, Swignle (1946) is totally silent. These species which have distinct specific characters and which have long been recognized by others cannot be quashed away on the assumption that they might have been of hybrid origin. When a particular form breeds true to type through several centuries, unaided by human agencies, what is the harm to recognize it as a variety or species as the case may be. Similar view has also been expressed by Hume (1941). It may be incidentally mentioned here that Swingle (1946) does not himself seem to incidentally mentioned here that Swingle (1946) does not himself seem to have adhered rigidly to the principle of not assigning specific rank to materials of hybrid origin as is evident by the fact that *Citrus paradisi* (grapefruit) which is suspected by him to have originated from the pummelo(*C.grandis)* has been given by him a separate specific status other than *C.grandis.* This kind of treatment of assigning specific or varietal names to fruits of unknown origin,persistent through ages, is not altogether uncommon in systematic botany for proper identification. In regard to varietal delineation, Webber's method has generally been adopted.

With the above idea in view the pattern of classification for Assam Citrus has been modelled and each and every collection has been given its rightful taxonomic status as far as possible. From the materials under study, sixteen different species excluding a new one and fifty two varieties occuring within the same have been identified and described fully. A few indubitable hybrids, presumably originated by natural cross-pollination, could not be accommodated within the conventional groupings, nor could they be classed separately as new species or varieties because certain prominent characters of some well known species are apparently hidden in them in such a way that the very external looks of the fruits and leaves suggest that they are of hybrid origin. Besides, these are of rare occurrence and have hardly any horticultural importance. These variants which could not be fitted in the general scheme of classification, have also been described fully and presented in the closing chapter, under the caption 'Probable Hybrids'.

With regard to the new species it may be stated that this cannot be fitted into any other known species nor can it be doubted for a hybrid as it has distinct specific characters and it breeds true to type and is widely distributed.

From the foregoing it will be seen that the primary basis of our work on classification is empirical. Not a single species or variety has been accepted unless it satisfies the taxonomical requirements very fully. Besides, no form has been given a varietal status unless it has been found to be well distributed in the province. The varietal names have been given along with the specific names for easy and proper identification. This is in accordance with the International Rules of Botanical Nomenclature (Article 35, page 13, *Brittonio,* Vol. 6(1), April, 1947) which permits any cultivated varieties or forms of a species to be given a varietal name following the name of the species, for example in our case–*C.reticulata c.sooh-niamtra, C.meidca c.bina-jara, C.jambhiri c.soh-myinidong* and so on.

## Acknowledgement

The authors place on record their grateful thanks to the Indian Council of Agricultural Research, Delhi and the Government of

Assam for jointly financing the Citrus Fruit Research Scheme, at Burnihat, Assam under which this piece of investigation was carried out. Thanks are also due to Dr H.K.Nandi and Dr S.D.Chaudhury successive Economic Botanists, Assam for their valuable suggestions at the time of investigation in its early stage, to Mr. Mawsing, Assistant Deputy Director of Agriculture, Hills, Assam and to numerous Forest and Political Officers of Assam for their help in collection of some wild specimens, to Professor H.Goswami, Department of Chemistry, Cotton College, Gauhati for undertaking chemical analysis of differnet fruit samples, to Mr. S.Chowdhury, Mycologist, Assam for helping us with some reference books, to Father Thurmes of St. Xavier's College,Calcutta for Latin translation of the botanical description of a new Citrus fruit and specially to Dr. K.Biswas, Superintendent and Mr. Narayanaswami of Royal Botanic Garden, Calcutta and to Mr. B.B.Shyam, Forest Herbarium, Shillong for rendering valuable help in identification of some specimens.

We are indebted to Dr S.K.Mitra formerly Director of Agriculture, Assam and to Mr. R.C.Woodford, O.B.E., the present Director of Agriculture, Assam but for whose encouragment and help in various ways this protracted investigation could have been impossible.

The authors are also specially thankful to Mr. A.K.Roy of the Citrus Fruit Research Station, Burnihat for preparing various coloured illustrations and photographs furnished in the text.

To the numerous colleagues and officers of the Department of Agriculture, Assam who have helped us in various ways our thanks are lastly due.

## Citrus Linn.

Trees or shrubs with sharp spines; *leaves* 1-foliate, alternate coriaceous or chartaceous, punctate with pellucid glands, aromatic; *petiole* often winged, articulated with leaf blade; *flowers* solitary or in small cymes or racemes, small to large, bi-sexual and staminate, sweet scented; *calyx* cup-shaped, 4-5 lobed; *petals* 4-8, usually 5, coriaceous, linear, gland-dotted imbricate, white or tinged; *stames* 20-60, *filaments* usually polyadelphous or free flattined at the base; *ovary* many-celled, 8-18, usually 10-14, style terete, deciduous, ovules in two series in each cell; *fruits* (berry) a hesperidum with segments containing seeds near the ventral side, remaining space of carpel filled up with stalked fusiform pulp-vesicles containing sweet or acid juice; *rind* leathery, foveolated with oil glands, turning yellow or orange red at full maturity; *seeds* none to many, obovoid to flattened obovoid, containing embryo, some times two or more in each seed, either white or green.

**Key to The Species of Citrus**

A. Pulp-vesicles free from acrid oil-droplets; petiole not or nearly winged; flowers large or small, white or tinged; stamens polyadelphous (sub-genus *Eucitrus)*.

B. Petiole absolutely wingless, not articulated; flowers tinged, staminate or 2-sexual; fruits large to very large, rind very thick, hard, sweet,` ........................................ (1) *C.medica.*

BB. Petiole with rudimentary wing or almost wingless, articulated.

C. Flowering all the year round, tinged, 2-sexual; fruits medium-sized (4-9 cm. diam.), oblong, broadly mamillate adherence of carpel strong, pulp usually white, sour or sweet; rind adherence strong. ......................... (2) *C.limon.*

CC. Flowering once annually, tinged, 2-sexual; fruit medium-sized (4.5-9.5 cm. diam.), very warty and bumpy, ovate-oblong; carpel easily separable; pulp yellowish white, sour or sweet; rind adherence medium to none. (3) *C.jambhiri.*

D. Fruit large (8-12 cm. diam.); rind thick, spongy, sweet; pulpvesicle orange tinged, intensely sour ....... (4) *C.karna.*

E. Rind very loosely attached; segments very easily separable; pulp-vesicles orange-tinged, usually sweet, rarely sour; petiole with rudimentary wing or wingless; flowers solitary, pure white.

F. Seeds small, clavate with fin-like projection of testa, not fattened, chalazal cap brown, cotyledon green, polyembryonic; fruit medium-sized (7-8 cm. diam.), peel orange coloured. ........................................... (7) *C.reticulata.*

FF. Seeds large, few, flattened, orbicular; leaf caudate; petiole short, rudimentary wing; pulp-vesicles slimy; fruit very smal (2.5-4 cm. diam.), peel deep orange red or scarlet. ... .................................................................... (9) *C.indica.*

G. Leaf yellowish green; petiole wingless, short; flowers large pure white, pulp-vesicles white, very sweet (acidless), aroma distinct; seeds few, chalazal cap yellow ochre; fruit medium-sized (6-7 cm. diam.), peel thin, light yellow. ..... ........................................................................ (6) *C.limetta.*

H. Fruit very large (12-14 cm. diam.); petiole with rudimentary wing; flowers white or slightly tinged; pulpvesicles acid, bitter principle not marked, melting, not easily separable, white or tinged. ................................ (13) *C.megaloxycarpa.*

BBB. Petiole with moderately large to spathulate wing, articulated.

K. Aroma of leaf distinct, (typical of bitter orange); petiole broadly winged; pulp-vesicles orange-tinged very sour, bitter; seeds cuniform; fruit medium-sized (6-8 cm. diam.), scarlet red .................................................. (11) *C.aurantium.*

KK. Aroma of leaf distinctly different from that of bitter orange; petiole narrowly winged; pulp-vesicles sweet, orange-tinged; fruit medium-sized (6-8 cm. diam.), peel orange, adherence strong ........................................ (10) *C.sinensis.*

KKK. Aroma of leaf strong, approaching eucalyptus smell; petiole obovate to spathulate, margin distinctly revolute; flowers all the year round, deep purple, petals 4-8; pulp-vesicles slender, very sour; seeds D-shaped, chalazal cap deep chocolate; fruits slightly rough to smooth, peel greenish yellow, medium-sized (7-9 cm. diam.), .............................. ................................................... (17) *C.assamensis.* Sp. nov.

L. Pulp-vesicles obovoid to very slightly ovate, peculiarly arranged arising from radial walls of the carpels, orange-tinged,sweet; seeds somewhat cuniform; fruit medium-sized (6.5-8 cm. diam.); rind medium thick, pebble surface, adherence with carpels slight or none. ........... (8) *C.nobilis*

LL. Pulp-vesicles slender, fusiform, greenish white, usually very acid; seeds very small; fruits very small (4-5 cm. diam.), peel very thin, greenish yellow; petiole obovate to spathulate; flowers all the year round. ............................. ...................................................................... (5) *C.aurantifolia.*

M. Fruit very large (10-18 cm. diam.), either globose, or oblate, or pyriform; petiole broadly winged, cordate or subcordate, blade pubescent underside; flowers large, pure white, racemes; pulp-vesicles sub-acid, bitter principle marked, easily separable, colour white, pink, or honey-yellow; seeds many, wedge shaped, rough, monoembryonic.............................................................................................. (12) *C.grandis.*

AA. Pulp-vesicle containing acrid oil-droplets; petiolar wing as large as the blade or larger than the blade;flowers medium to large, stamens usually free or cohering in bundle (sub-genus *Papeda).*

N. Leaves accuminate, or caudate; stamens coherent; pulp-vesicles globose to obovoid, white, sour, juice scanty; seeds many, very large, angled, chalazal cap brown, covering a wide area.................................................. (14) *C.ichangensis.*

NN. Leaves not caudate; seeds many, medium-sized, not angled, thin; fruit small to medium.......................... (15) *C.latipes.*

O. Flowers small to medium, tinged, stamens free, style very short.

P. Fruit medium to large (5-12 cm. diam. usually 5-6 cm. diam.), globose, smooth; petiole as large as or often larger than blade; pulp obovoid, arising from radial walls, sour. ............................................................................ (16) *C.macroptera.*

(In the preparation of the key to the species Swingle's key for the same has been basically copied by the writers.)

# Chapter 2

# SUBGENUS EUCITRUS

## The Citron (*Citrus medica* Linnaeus)

*Origin and name.* Roxburgh (1832) mentions that the real citron grows in the Garo Hills of Assam. Wight (1840-50) reports that the citron has been found to grow wild in the Nilgiri Hills. Watt (1889) considers it to be a native of Kumaon, while A.de Candolle (1886, 1912) says that the species citron is of Indian origin. Similar observation has also been made by Lindly and Moore (1866). In the *Flora of British India,* Hooker (1872) states that the citron is commonly found in the warm valleys along the foot hills of the Himalayas from Garhwal to Sikkim ascending to 4,000 feet in the Khasi Hills and in the Garo Hills and also in the Western Ghat and Satpura range in Madhya Pradesh.

Bonavia (1890) while discussing the origin of the *Citrus* expresses that it is doubtful whether *Citrus medica* (citron) is indigenous to india, but at the same time he opines that if India ever possessed an indigenous citrus it is probably the citron. Swingle (1946), however, entertains a different view and does not consider citron to be of Indian origin. He assumes that the probable home of the citron is in the region between Hadhramaut and Oman.

In the *Account of Assam* Robinson (1841) has mentioned of the citron under different Assamese names which still hold good today. Writing on the *Flora and Vegetable products of Assam* Captain Lowther (1857) said that at Saikhowa (Upper Assam) he had seen some citron plants growing almost wild on a barren island of nearly pure sand and that its creeper like trailing branches were heavily laden with ripe perfumed citron.

Going further into the past one finds mention of the citron in the old literature and arts of India, *e.g.*, the *Vajasaneyi Samhita* which is supposed to have been written in a period anterior 800 B.C., the

*Charka Samhita* written in A.D. 100,and the world famous *Ajanta* paintings, completed during the period from 200 B.C. to 700 A.D. Even the old Chinese literature mentions citron and fingered citron (*C.medica* var. *sarcodactylis)* the *fo-shu-kan*, meaning Buddha's hand. The association of the fruit with the name of the Holdy Buddha leads one to believe that it is of Indian origin. The rival claim of Media and Persia as an original home of citron is not sustainable except on the basis of conjecture. Taking all facts into consideration it seems probable that the citron is indigenous to India.

The true citron is called *bira-jora* or *bakol-khowa-tenga (bakol* means rind, *khowa* to eat, and *tenga* stands for sourness) in Assam Valley plains, *jaarajamir* in Sylhet and Cachar, *soh-madheh* or *soh-manong* in the Khasi Hills, *lume-han-thor* in the Mikir Hills, *haijange* in Manipur, *naya-changney* in the Daffla Hills, *sutrung* in Bengal, *bhimra* in Nepal, *mokri* in Madras, *mokari* in Punjab, *mohalung* in Bombay, *madh-kunkur or madh kakree* in the Uttar Pradesh, *maulung* in Goa and *siderun or natterun* in Ceylon. All these names of citron in India go, as well, to show that it is probably indigenous to India, otherwise there would not have been so many names for *Citrus medica.*

*Nomenclature.* Of all the Citrus fruits citron is very widely known and is the least disputed for its nomenclature. Like most other Citrus fruits citron has been recognized by various scientific names, *viz. Citrus tuberosa* Mill., *Gord. Dict.*, 8th ed. (1768); *C.odorata* Roussel (1786); *C.fragrans* Salisb. (1796); *C.cedra* Link (1831); *C.cedratus* Raf. (1838), *C.crassa* Hassak (1844), etc. But all modern authors unanimously have accepted the oldest Linnaean name *Citrus medica* for citron.

Webber (1946) segregated citron into two distinct groups according to the characters of sweetness and acidity of pulp. In Assam similar grouping can also be done. The different varieties of citron that are found growing scattered in villages as well as in the primeval forests of the province can be placed conveniently into three distinct varieties to represent correctly the citrons of Assam. Incidentally, it may be stated that the fingered citron *(C.medica,* var. *sarcodactylis)* has not been met with anywhere in Assam. For different Indian varieties of citron a reference may be made to Bonavia's (1890) *Oranges and lemons of India and Ceylon* (Illustration) Plate Nos. CXLII to CLVIII.

All the representative varieties of the citron are keyed below.

## Key to Varieties of The Citron

A. Pulp-vesicles very acid, juice scanty.

B. A small tree or shrub, height 8-12 ft.; straggling branches; thorny foliage open leaf ovate-oblong or elliptical, apex obtuse, margin usually serrate; petiole short, wingless, not articulated; young shoot slightly purple tinged.

C. Fruit very large, size 20-40cm. ht., 14-28 cm. diam.; wt. 8-20 lb.; shape ovate-oblong; surface slightly smooth to sub-warty; rind em., earrot like consistence, sweet.

$V_3$...............*Mitha-jora.*

CC. Fruit medium large, size 15-19 cm. ht., 10-12 cm. diam., wt. 2-3 lb., shape ovate-oblong; surface slightly smooth to sub-warty; rind thickness 1.8-3 cm., carrot like consistence, sweet.

$V_2$..............*Soh-manong.*

AA. Pulp-vesicles sweet, juice scanty.

B. Fruit medium large, size 18-22 cm. ht., 10-15 cm. diam., wt. 3-4 lb.; shape ovate-oblong or ellipsoid, surface almost smooth to very slightly sub-warty; rind thickness 4-6 cm., carrot like consistence, sweet.

$V_3$..............*Mitha-jora.*

### *Bira-jora*

*Plant characters.* A small tree or a shrub, 8-12 ft. long tendency to lodge; pendulous, sagging branches, branching very much low from the trunk; thorny, 2-4cm. long, stout; foliage open; fruiting season mainly in September to December, also shy crop all the year round.

*Leaf characters.* Leaf usually large, ovate-oblong or elliptic-ovate, size 14-18 cm. long, 7-9.5 cm. broad, medium coriaceous, colour light green, articulation obscure or none, apex either obtuse or slightly acute, margin serrate to slightly crenate; petiole wingless, very short, 0.8-1 cm. long; young buds light purple tinged.

*Flower characters.* Bisexual and staminate, percentage of staminate flowers more than the pistillate, glomerate, 4-12, cymes; pedicel short, 4-6 mm. long, peduncle short, 0.5-1 cm. long, glabrous;

calqx purple tinged, dotted with pellucid glands sepals 4-5, cupular, coriaceous; corolla imbricate, petals 4-5, size 2.4-3 cm. long, 7-9 mm. broad, colour purple outside, shape oblong to oblanceolate, dotted with pellucid glands; stamens 30-58 or occasionally more, anther 6-8 mm. long, filaments white, slightly pubescent, polyadelphous, 1 or 2 free,unequal in length, 0.9-2cm. long; pistil club-shaped, style terete, light purple or white, 1-1.5 cm. long including stigma,stigma pinkish; ovary 7-8 mm. long, 4-5 mm. diam; flowering season mainly in February and March, also all the year round.

***Bira-jora (C. medica)***

*External fruit characters;* Shape oblong to ellipsoid; size variable according to season of fruiting, normal size 20-40 cm.long,14-28 cm. broad; weight 8-20 lb.; D/H index av. 0.69; colour lemon yellow when full ripe; surface somewhat smooth to sub-warty, waxy, pitted or foveolated; base slightly depressed, almost obtuse; apex broadly mamillate, sometimes obtuse; areole indistinct, areolar area slightly wrinkled or ribbed, stylar scar even, style occassionally partially persistent.

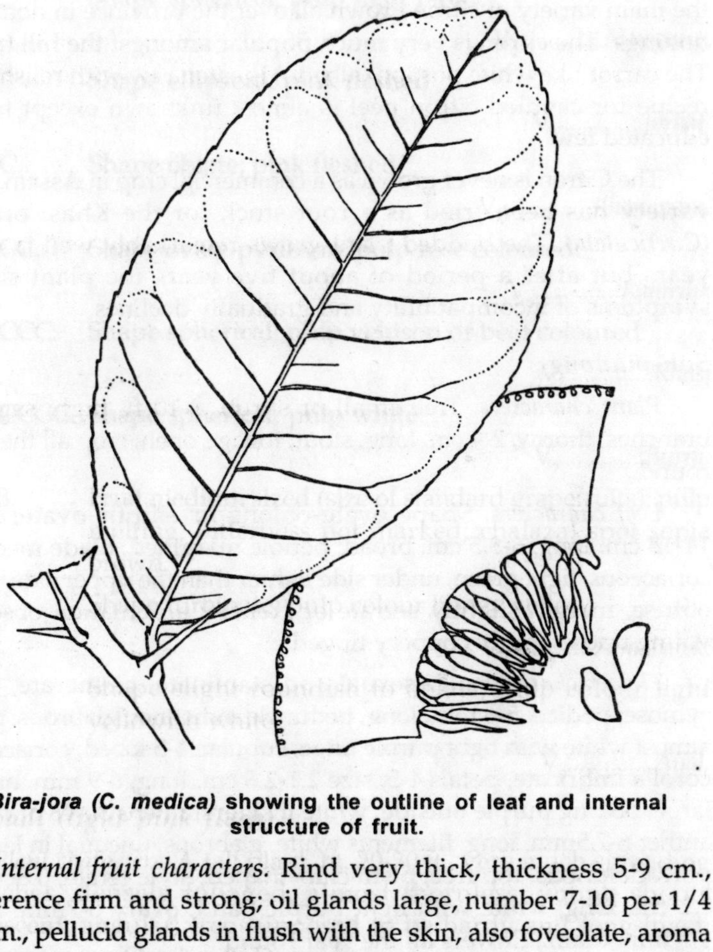

**Bira-jora (C. medica) showing the outline of leaf and internal structure of fruit.**

*Internal fruit characters.* Rind very thick, thickness 5-9 cm., adherence firm and strong; oil glands large, number 7-10 per 1/4 sq. cm., pellucid glands in flush with the skin, also foveolate, aroma pleasant; mesocarp (albedo) white, texture hard, carrot-like consistence, sweet; axis usually semi-hollow or occasionally solid in small fruits; segments 10-12; rag scanty; pulp-vesicles white, texture somewhat coarse, vesicles loosely packed, not melting apex both acute and obtuse, usually 1.5-2 cm. long; juice scanty, occasionally 1.5 oz. per fruit, taste very sour; seeds many, 30-40 per fruit, rarely 60 or more, medium sized, usually 1.1-1.2 cm. long, 6-7 mm. broad, 4-5 mm. thick, shape usually obovate, surface almost smooth, minute anastomosing lines on the edges, testa cream white, tegmen light brown, chalazal spot purple or deep chocolate, cotyledon white.

The *Bira-jora* of Assam is the real citron of commerce. This is the main variety of citron grown all over the province in domestic hodings. The citron is very much popular amongst the hill tribes. The carrot like white portion (albedo) is eaten raw with relish. The recipe for candied citron peel is almost unknown except to the educated few.

The Citron is never grown as a commercial crop in Assam. This variety has been tried as a root stock for the Khasi orange (*C.reticulata).* The budded plant grows remarkably well in early years but after a period of about five years the plant shows symptoms of incompatibility and gradually declines.

### *Soh-manong*

*Plant characters.* Tree small or shrub, 8-10 ft. high; sagging branches; thorny, 2-4 cm. long, stout; foliage open; crop all the year round.

*Leaf characters.* Shape ovate-oblong or elliptic-ovate; sizes 14-18 cm. long, 6-8.5 cm. broad; petiole margined; blade medium coriaceous, light green, under side lighter than the upper side, apex obtuse, margin slightly serrate or crenate; articulation obscure; young buds slightly coppery tinged.

*Flower characters.* Bi-sexual and staminate, glomerate, 4-12, cymose; pedicel 5-6 mm. long, peduncle indistinct, glabrous; calyx almost white with light purple tinge, cupular, 4-5 lobed, coriaceous; corolla imbricate, petals 4-5, size 2.1-2.8 cm. long, 6-9 mm. broad, large, oblong purple outside, white inside; stamens 30-48 or more, anther 6-7.5mm. long, filaments white, glabrous, unequal in length, polyadelphous, top free; pistil club-shaped; style terete, 0.8cm. to 1.2 cm. long, white with light purple band; ovary 6-7mm. long, 4-5 mm. diam.; flowers all the year round.

*External fruit characters.* Shape oblong or ellipsoid,tapering at apex and somewhat obtuse at base; size variable according to season, size 15-19 cm. ht., 19-12 cm. diam; wt. 2-3 lb.; D/H index av. 0.64; colour lemon yellow; surface almost smooth to slightly sub-warty; base almost obtuse, basal area slightly depressed; calyx coriaceous, irregularly divided, tips obtuse or slightly acute; apex braodly mamillate or occasionally almost obtuse; areole indistinct; stylar scar even or occasionally depressed in a small apical pit, style occasionally partially persistent.

*Internal fruit characters.* Rind thick, thickness 1.8-3cm., adherence strong; oil glands number 6-10 per ¼ sq. cm., prominent, usually in flush with the skin or pitted, aroma pleasant; mesocarp (albedo) white, texture hard and compact, carrot like consistence, sweet; axis usually hollow, occasionally solid; segments 10-12; rag little; pulp-vesicles white, somewhat coarse, fusiform, 1-1.6 cm. long; juice scanty, rarely 1.5 oz. per fruit, very sour, flavour agreeable; seeds numerous, 24-35 or more, size medium, usually 0.9-1 cm. long, 0.4 cm. broad, 6-7 mm.thick, shape ovate-oblong to obovate, surface almost smooth, micropylar end attenuated and obtuse, testa cream white, tegmen light brown, chalazal spot purple or deep chocolate, cotyledon white.

This variety of citron is met with everywhere in the province. It bears a striking similarity in respect of different morphological characters, except probably for the colour of the chalazal spot, with the variety *Diamante* (Cedro liseio) as described by Webber (1946) in the *Citrus Industry,* Vol. I, Pp. 635. This has also been tried as a rootstock for the Khasi orange (mandarin) and found to be incompatible.

### *Mitha-jora*

*Plant characters.* A small tree or shrub, 8-10 ft. high, tendency to lodge; straggling branches; thorny, 2-3.5 cm. long; foliage open, fruiting season all the year round, main crop in September to December.

*Leaf characters.* Shape oblong or elliptic-ovate, large, size 12-16 cm. long 6.5-8 cm.broad,petiole wingless, very short, 1 cm. long,articulation obscure, leaf blade light green,chartaceous, veins prominent on underside, apex somewhat obtuse or slightly acute, margin crenate or occasionally slightly serrate; young shoots very slightly coppery tinged.

*Flower characters.* Bi-sexual and staminate, glomerate (clustered) 3-12, cymes; pedicel 4-5 mm. long, peduncle very short, 3-5 mm.long, glabrous; calyx pruple tinged, lobes 4-5, cupular, coriaceous; corolla imbricate, petals 4-5, colour very light purple tinged outside, white inside, size 2-2.8 cm. long, 6-7 mm. broad, shape oblong; stamens 30-45 or more, anther 6-7 mm. long filaments white, minutely pubescent, 1-1.5 cm. long; polyadelphous, pistil clubshaped, style terete, 1-1.2 cm. long; flowers all the year round, mainly in February and March.

*External fruit characters.* Shape ovate-oblong or ellipsoid, size large, usually 18-22 cm. ht., 10-15.5 cm. diam; weight 3-4 lb.; D/H index av. 0.68; colour lemon yellow; base almost obtuse, basal area slightly depressed and very minutely rugose surface slightly smooth or very slightly sub-warty, waxy, pitted, glands also in flush with the skin; calyx coriaceous, division irregular, tip obtuse or slightly acute; apex broadly mamillate or occasionally rounded, off season fruits usually deformed and small; areole indistinct, areolar area slightly rough; stylar scar even or very slightly elevated,style usually not persistent.

*Internal fruit characters.* Rind thick, thickness 4-6 cm. adherence strong; oil glands distinct, number 6-10 per ¼ sq. cm., shape globose or elliptical, aroma pleasant; mesocrap (albedo) white, hard and compact, carrot-like consistence, sweet; axis usually hollow; segments 10-12; pulp-vesicles white, texture coarse, fusiform, very loosely packed, 1-1.6 cm. long; juice scanty, sweet or insipid sweet (lack of acidity); seeds numerous, 28-38 or more per fruit, size usually 1-1.1 cm. long, 6-7 mm. broad, 5.6-6 mm. thick, shape ovate-oblong, irregular, surface smooth, micropylar end atenuated, obtuse or acute, testa cream white, tegmen light brown, chalazal spot chocolate, cotyledon white.

The botanical descriptions of this variety seem to conform to those of the Corsican, as described by Webber (1946) except for the colour of the tegmen and chalazal spot which are said to be cream and yellow buff respectively in the case of the Corsican.

This variety has also been tried as a rootstock for the Khasi orange (mandarin) and Valencia Late but it has been found to be incompatible. It has no commercial importance either.

### The Lemons (*Citrus limon* (Linn.) Burmann)

*Origin and name.* The nativity of the lemon is a matter of conjucture. That South Eastern Asia as its natural home seems to be agreed upon by most of the authors; but whether India can exclusively claim it as its own is disputed. As eary as 950 A.D. the Persian geographer Istakhri had noticed the fruit in the Punjab under the name *limun* and again an Arabic writer Idrisi had recognized a similar fruit in the Punjab in A.D.1175 under a slightly changed name *limuna*, Laufer (1934) considered that the lemon is an introduced variety in China as the word *li-mung* used in the earliest Chinese literature for referring to the lemon is altogether foreign to Chinese language. Even in the present day, the lemon is variously called in different parts of India as *nimboo, nimu, nemu, lebu, nebu, limbu, limboo*, etc., which again has a striking phonetic resemblance to the word *limun, limuna* and *li-mung*, referred to above. It is not unlikely that the English word lemon has been derived by modification from the earliest Indian word *limun* although Bonavia considers that it has probably originated from the Malay word *lemoen*. All these phonetic resemblances among various names are indicative of a common origin for lemon, as corruption of words in the evolution of different languages is not uncommon. Tolkowsky (1938) says 'If in China, Japan and Indo-China oranges and shaddocks are the Citrus fruits commonly grown, in India it is the citron, the lemon, the lime that predominate; this is true for wild varieties as well as for those that are cultivated.' Glidden (1937) after going through various historical records of derivative meanings of different names used for lemons in different countries considers the lemons to be a native of eastern Himalayan region which obviously includes places like Assam where lemons are so numerous. It is therefore, not unnatural that the primeval forest of Assam where lemons of various kinds are found even today to grow in a wild state,undistributed by human hands, might be a probable home of this *Citrus* group.

*Nomenclature.* Lemon has been differently named by different taxonomists from time to time as recorded by Swingle (1946) and others:*C.medica* var. *limon*, Linu., *Sp. Pl.* 2:782 (1753); *C.limon*(Linn.) Burmann.*f.Fl. Indi.*, pp.173(1768); *Limon vulgaris* Mill (1768); *C.limonum (vulgaris)*. Risso (1813); *C.medica-limonum*, Brandis *For. Fl.* (1874); *C.aurantium* var. *limonum;* W. and A. Prodr. 98. *C.limonum*, Wall. Cat. 6389; Dalz. and Gibs. (1861) *Bomb.Fl. Suppl.* 13; *C.medica*, Willd., according to Roxburgh *Fl. Ind.* iii 392 (1832); *C.medica* var. *limonum*, Hooker, *Fl. Brit. Ind.* 1:515 (1872-1875) *Citrus limonia* Osback, Swingle (1919) in Bailey's *Stand.Cycl. Hort.* Vol. 2,pp. 781 and others.

The oldest name *Citrus limon* (Linn.) Burmann has been accepted of late for true sour lemons of commerce withiout any contradiction (Tanaka, 1937 a; Swingle, and Webber, 1946).

Like many other citrus species the lemon presents a numbr of diverse forms slightly varying from each other. There are forms met with which show a close fusion of some characters of both true lemon and its closely allied species the citron and it is these intermediate forms which add to confusion in proper grouping of lemon. The occurrence of this phenomenon probably led Linnaeus to consider lemon to be a variety of *Citrus medica.* Swingle (1946) also suspected lemon to be a hybrid originating as a satellite species of the citron though he conceded to it a specific rank as it reproduces itself from seeds with small variations when self pollinated. Bonavia (1890) clubbed such intermediate forms as *Citron-lemon* or *lemon-citron,* according to the predominance of the characters of the citron or of the lemon as the case may be and placed these under the name *C.medica* var. *bajoura,* separating them from the true lemons. In course of the studies on lemons and citrons of Assam it has been possible to garner many intermediate forms such as *pati-lebu, shasni-lebu, soh-madrit, jora-tenga,* etc. These are a few of the many forms to illustrate the range of variation within a common group. On close and careful examinations, these intermediate forms that have been considered to be akin to and indistinguishable from the true lemon have been placed under the lemon group. These forms as they exist now can never be separated clearly from the true lemon and reshuffling is inconceivable in any scheme of classification for fear of adding to further confusion. The only reshuffling of the lemon group that can justifiably be done for the sake of clarity is the separation of the so called rough lemon or *jambhiri* group which had hitherto, been classified by many under lemon. This has been dealt with fully later under *Citrus jambhiri.*

The varieties of the lemons of Assam are divisible into two distinct groups, easily separable by their acidity and sweetness of pulp-vesicles as suggested by Webber (1946). The position of the sweet lemon, in a group predominated by the acid lemon which universally carries the sense of sourness and not sweetness, is apparently anomalous. But as many other citrus groups have both acid and sweet forms, *e.g.,* in citron, lime and mandarin this anomaly cannot be considered as a serious bar in grouping sweet and acid lemon together. The *pani-jamir* which is the solitary representative of the sweet lemon in Assam bears a close resemblance to lemons by its slightly drooping habit of plants, shape of the fruit, colour of the chalazal spot and the purple pigmentation of the flower, though ephemeral. The distinctive differences of the *pani-jamir* with the true sour lemons lie primarily in its sweetness of pulp and annual flowering and fruiting habit. Sweet lemon was designated as *C.lumia* by Risso (1818).

## Key to Varieties of the Lemon

A. Pulp-vesicles intensely sour (acid lemon group).

B. Flowering all the year round, purple tinged.

C. Tree very small or bushy, ht. 4-6 ft.; leaf ob-lanceolate, size decidedly smaller than those mentioned in the lemon group, arranged almost spirally; internode very short.

D. Fruit very small, 3.3-4 cm. ht., 3.5-4.3 cm. diam., surface smooth, section showing through the skin; apex obtuse or very slightly nippled, stylar scar slightly protruded, style partially persistent; juice scanty; very seedy.

$V_1$............*Naya-changney.*

CC. Small tree to shrub, ht. 6-8 ft., somewhat staggering habit; branches spreading; moderately foliaged; young shoots slight coppery tanged; thorny; leaf pale green; shape oblong; apex usually acute; petiole indistinct or very short, articulation obscure.

D. Fruit oblong; size 10-14 cm. ht., 5-7 cm. diam., narrowed towards base, broader towards apex, low-nippled; surface very smooth and glossy; flavour mild, typical lemon aroma; seeds few to almost seedless, chalazal spot deep purple.

$V_2$............*Assam-lemon.*

DD. Fruit lemon-shaped or ovate-oblong; size 9-12 cm. ht., 7.5-8.2 cm. diam., surface smooth and glossy; apex with very small and somewhat pointed low-nipple; flavour strong, distinct; seeds many.

$V_3$............*Pati-lebu.*

DDD. Fruit oblong, narrowed towards base and apex; size 12-15.5 cm. ht., 7-8.2 cm. diam.; apex broad and long nippled; surface subwarty, dull.

$V_4$.............*Jora-tenga.*

DDDD. Fruit ovate-oblong, tapering towards base and apex; size 12-13.6 cm. ht., 8.5-9.4 cm. diam.; broadly nippled; surface warty and pebbled, dull.

$V_5$..........*Godha-pati-lebu.*

DDDDD. Fruit oblong or ovate; size 9.5-11 cm. ht., 6.5-7.5 cm. diam.; apex abruptly protruded into braod and long nipple; surface rugose, puffing; flavour mild, chalazal spot light purple.

$V_6$.............*Kuta-jamuri.*

CCC. Tree medium tall, ht. 8-10 ft., staggering branches; moderately foliaged; young shoots coppery tinged; leaf slightly oblong to elliptical, apex obtuse, dark green, crushed leaf aroma strong approaching cardamom, petiole with rudimentary wing, articulatin on upper surface only.

D. Fruit oblong, tapering towards base, broader towards apex; size 12-14.5 cm. ht., 7-7.5 cm. diam., nippled; surface almost smooth, slightly rough towards base; necked; flavour distinct, approaching cardamom; seeds few, chalazal spot deep purple.

$V_7$.............*Elachi-lebu.*

BB. Flowering once annually, purple tinged.

C. Tree medium to tall, ht. 10-25 ft., spreading, moderately dense foliage; thorny; young shoot tinged; blade ovate-oblong, light green, petiole distinct, articulated.

D. Fruit lemon-shaped, size 11.5-14 cm. ht., 8-11 cm. diam.; surface rugose to slightly warty, section showing through rind; apex broadly mamillate, flavour distinct; seeds few, size large, chalazal spot purple.

$V_8$............*Soh-long.*

DD. Fruit nearly spherical, size 7-8.2 cm. ht., 7.5-7.8 cm. diam.; surface very smooth, glossy, light lemon yellow colour, apex slightly nippled; seeds few, chalazal spot purple.

$V_9$............*Soh-synteng.*

AA. Pulp sweet or insipid sweet (sweet lemon)

B. Fruit lemon-shaped, size 13-16 cm. ht., 8.5-11.8 cm. diam.; surface almost smooth or rugose, section showing through skin; apex broadly mamillate, flavour distinct, assisting diagnosis; seeds many, chalazal spot deep purple, cotyledon very slightly greenish white.

$V_{10}$............*Pani-jamir.*

## *Naya-Changney*

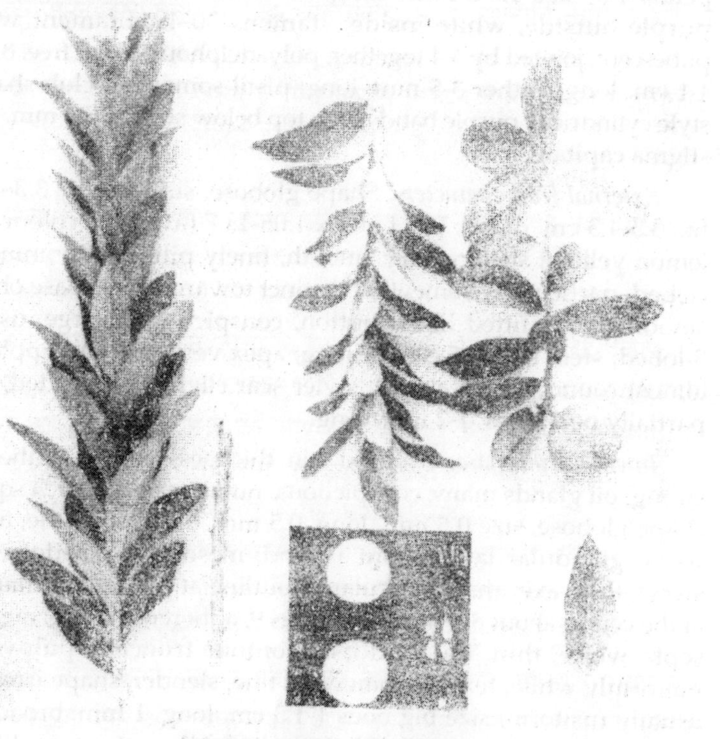

**Naya-changney (*C. limon*)**

*Plant characters.* A small bushy plant or a shrub, 4-6 ft. high; moderately dense foliage; very short internodes, 1-1.5 cm. long; spinous, small or rudimentary; shoots slightly coppery tinged, more or less terete; fruiting season mainly September to December, shy crops all the year round.

*Leaf characters.* Leaf small, medium coriaceous, petiole indistinct, 1-4mm. long, articulation obscure, shape ob-lanceolate, size 5-8.5cm. long, 1.8-3.1 cm. broad, margin crenate to bi-crenate, oil glands in every notch, apex attenuated, obtuse, emarginate, leaf spirally arranged with a tendency to crowd upwards; pellucid glands prominent, protruded over the surface.

*Flower characters.* Bi-sexual and staminate, glomerate (2-8), cymes; pedicel indistinct, sessile, peduncle very short, 2-6 mm. long, glabrous; calyx 4-5 lobed, tips tinged, cup-shaped; corolla imbricate, petals 4-5, size 1.8-2.1 cm. long, 5-7 mm. broad, oblong, colour purple outside, white inside; stamens 30-49, filament white, pubescent, jointed by 3-4 together, polyadelphous, 1 or 2 free, 8 mm. 1.1 cm. long, anther 3-5 mm. long; pistil somewhat club-shaped, style cylindrical, purple band at the top below stigma, 5-6 mm. long, stigma capitate.

*External fruit characters.* Shape globose, size usually 3.3-4 cm. ht., 3.5-4.3 cm. diam.; D/H index 1.05-1.07 (av. 1.02); colour light lemon yellow; surface very smooth, finely pitted, very minutely ribbed, particularly somewhat distinct towards apex, base obtuse, smooth; calyx fitted like a button, conspicuously large, usually 5-lobed; stem small, 5-8 mm. long; apex very slightly nippled or almost round; areole absent, styler scar clightly protruded, style partially persistent, 1-2 mm. long.

*Internal fruit characters.* Rind thin, thickness 3-4 mm., adherence strong; oil glands many, conspicuous, number 32-42 per ¼ sq. cm., shape globose, size 0.5 mm. long, 0.5 mm. broad, oil little, aroma weak, glandular layer about ⅓ rind; mesocarp (albedo) white, sweet, thin; axis almost circular in outline at the centre. diameter at the centre about 8-9 mm.; segments 9, adherence stong; rag little; septa white, thin, tough, dorsal contour truncate; pulpvesicle uniformly white, texture somewhat fine, slender, shape irregular, usually fusiform, size big ones 1-1.3 cm. long, 1 mm. broad, tips usually acute, occasionally blunt; juice very scanty, very difficult to squeeze out, very sour; seeds numerous, 60-80 per fruit, 5-8 seeds in each carpel, size medium 0.8-1 cm. long, 5-6 mm. broad, 3-4 mm. thick, shape almost orbicular to ovate with gradual tapering towards micropylar end, surface smooth, sometimes with fine anastomosing lines on one side, micropylar end somewhat blunt, testa white, tegmen deep brown, chalazal cap purple, cotyledon white.

This is really a wild lemon found to grow at an elevation of about 3,000 4,000 ft. in Daffla Hills. Daffla tribe called it *naya-changney*. The fruit is full of seeds with little juice in it. Its peculiar arrangement of leaf is distinctly different from that of the common citrus varieties. The fruits have hardly any commercial importance.

## *Assam-lemon*

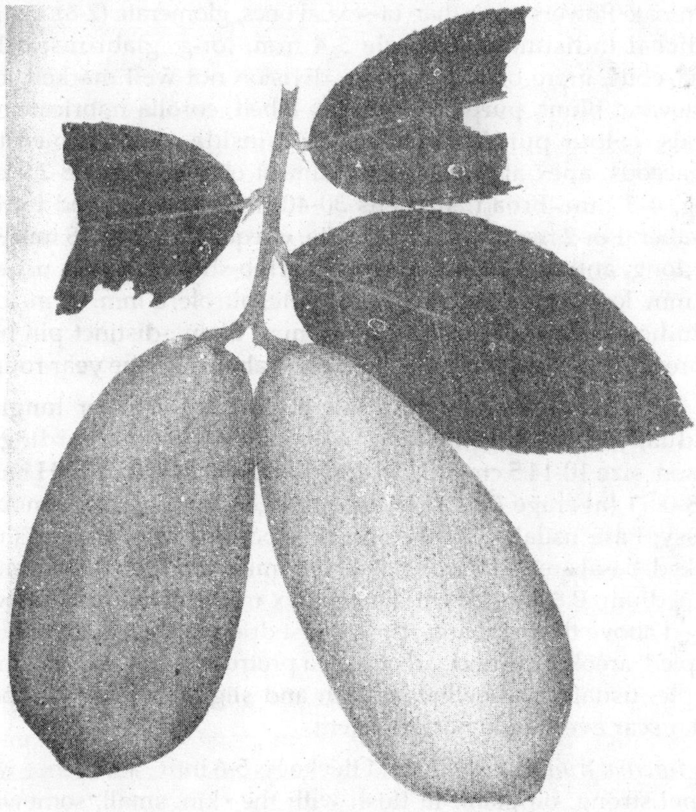

**Assam Lemon (*C. limon*)**

*Plant characters.* A small tree or shrub, 6-8 ft. high, well branched, straggling; moderately dense foliage; thorny, small, slender, usually 6 mm.-1.5 cm. long, occasionally 2-3 cm. long or more; fruiting season almost throughout the year, main crop July-September, prolific bearer.

*Leaf characters.* Leaf medium sized, 8-10.5 cm. long, 4.5-5.5 cm. broad; rarely 11-15 cm. long, 5.5-7.5 cm. broad; ovate-oblong, narrowed towards the apex, apex usually acute or slightly obtuse, margin serrate to crenate, leaf green upper side, pale green beneath, slightly articulated; petiole wingless, 6-8 mm. long; twigs very slighty purple tinged.

*Flower characters.* Both bi-sexual and staminate, proportion of staminate flowers more than bi-sexual ones, glomerate (2-8), cymes; pediceal indistinct; peduncle 3-4 mm. long, glabrous; calyx coriaceous, more or less cupular, division not well marked, tips somewhat blunt, purple tinged, 4-5 lobed; corolla imbricate, 4-5 petals, colour purple outside, white inside, oblong, medium coriaceous, apex attenuated and almost obtuse, size 1.8-2.9 cm. long, 4-8 mm. broad; stamens 30-40, filaments jointed by 2-3 together, 1 or 2 free, pubescent, white, unequal in length, 8 mm.-1.5 cm. long, anther 5-6 mm. long; pistil club-shaped; ovary usually 7-8 mm. long, 3-4 mm. diam; style terete, purple, 8 mm.-1 cm. long including stigma, stigma large, 4-5 mm. diam., distinct pin hole depression at the top; flowering season almost all the year round.

*External fruit characters.* Shape oblong to rather longish, gradually narrowed towards base; size variable according to season, size 10-14.5 cm. ht., 5-8 cm. diam., wt. 6-11 oz.; D/H index 0.48-0.53 (average 0.51); colour lemon yellow; surface smooth; glossy; base usually evenly rounded, occasionally slightly short necked, basal area narrow, even or very minutely ribbed; stem short to medium, 0.8 cm.-1.5 cm. long; calyx medium coriaceous, well raised above the basal area, tips almost dried up; apex slightly low nippled; areole indistinct, areolar area protruded into low and short nipple, usually somewhat smooth and slightly radially ribbed; stylar scar even, style not persistent.

*Internal fruit characters.* Rind thickness 5-6 mm., adherence with carpel strong; oil gland in flush with the skin, small, somewhat globose, 1 mm. diam. or slightly less, oil glands number 30-36 per ¼ sq. cm., aroma distinctly lemon; mesocarp (albedo) white sweet, texture hard; axis usually holow or semihollow, rarely solid, more or less uniform from base to apex, 6 mm. to 1 cm. diam. at the centre; segments usually 9-12, size of carpel generally 9 cm. long. 2.3 cm. broad, 1.8 cm. thick; septa white, distinct, chartaceous, dorsal contour convex to slightly truncate; pulp-vesicle crystal white, more or less acicular or needle-shaped, tip almost acute, 1.2-1.6 cm. long,

1 mm. diam., closely packed; juice abundant, wt. 2-5.5 oz., very sour, flavour good; seeds none or several, 0-12 number per fruit, usually seedless, ovate-oblong, smooth, micropylar end more or less obtuse, size 0.9-1.1 cm. long, 6-7 mm. broad, 4-5 mm. thick, testa cream white, chalaza spot deep purple, tegmen brown, cotyledon light greenish white or almost white.

**Assam lemon (*C. limon*)**

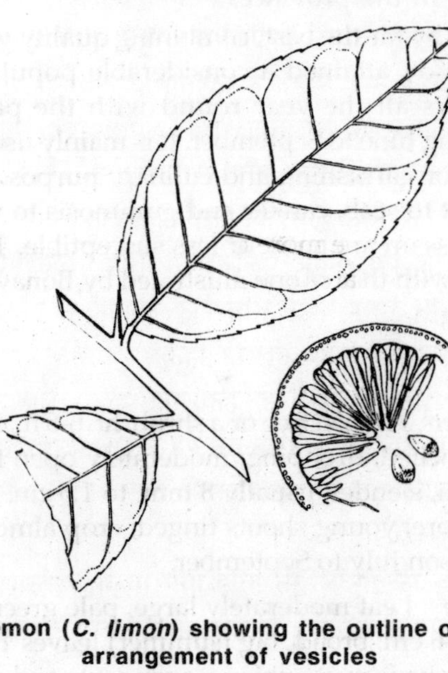

**Assam lemon (*C. limon*) showing the outline of leaf and arrangement of vesicles**

This variety has originated as a chance seedling raised in the Citrus Station, Burnihat from the progenies of a variety collected under the name *china-kaghi* from the village Hashara in the district of Sibsagar, Assam. The fruit of this chance seedling has been found to be strikingly superior to its progenitor in respect of yield and quality. This has been propagated by vegetative means as a clonal horticultural variety and named as Assam-lemon.

**Pati-Lebu (*C. limon*)**

In Assam, lemons are grown in almost every home for catering to domestic needs and it is only recently that large scale plantations on commercial lines are being started. Of all the lemon varieties grown in the province, Assam-lemon is by far the best, combining quality with yield. The variety has already attained a considerable popularity with the growers. It bears all the year round with the peak period of production during June to September. It is mainly used in the green matured stage for garnishing and culinary purposes. The variety is fairly resistant to scab, canker and gummosis to which most of the lemons of Assam are more or less susceptible. The fruit bears some similarity with that of one illustrated by Bonavia (1890) Plate No. CLXVIII, Fig. a.

### *Pati-lebu*

*Plant characters.* A small tree or a shrub, ht. 6-8 ft., irregular habit; moderately branched, drooping; moderately open foliage; thorns small to medium, slender, usually 8 mm. to 1.5 cm. long, big ones 3 cm. long or more; young shoots tinged; crop almost all the year round, peak season July to September.

*Leaf characters.* Leaf moderately large, pale green, size usually 9-12 cm. long, 4-6 cm. broad, big (summer) leaves 14-16 cm. long, 7- 8.5 cm. broad; shape ovate-oblong, apex acute and obtuse, margin serrate to slightly crenate; petiole wingless, short, 6-8 mm. long, minutely articulated.

*Flower characters.* Bi-sexual and staminate, glomerate (2-10), cymes; peduncle 2-3 cm. long; pedicel short, 4-5 mm. or rarely 1 cm. long, glabrous; calyx medium coriaceous, dotted with oil glands, sepal purple tinged, cupshaped; corolla purple outside and white inside, petals 4-5, size 1.8-2.6 cm. long, 6-7 mm. broad, oblong; stamens 30-45, anther 4-6 mm. long, filament polyadelphous, unequal in height, 0.8-1.5 cm. long, top free, pubescent; pistil more or less club-shaped, style cylindrical, 6-8 mm. long, 1 cm. long including stigma, stigma capitate with a pin hole depression at the top, ovary 4-6 mm. long, 3-4 mm. broad; flowering season all the year round, mainly in March and April.

*External fruit characters.* Shape obovate or ovate-oblong; size variable, normal 9.5-12 cm. ht., 7.5-8.2 cm. diam.; wt. 10-13 oz.; D/H index 0.68-0.77 (average 0.72); colour lemon yellow, smooth and glossy, very finely pitted; base usually evenly rounded, rarely short necked, basal area minutely rugose or smooth; stem short, 1-1.5 cm. long; apex nippled, areole indistinct, sometimes radially ribbed, areolar area protruded into small nipple, usualy more or less smooth, stylar scar even, style not persistent.

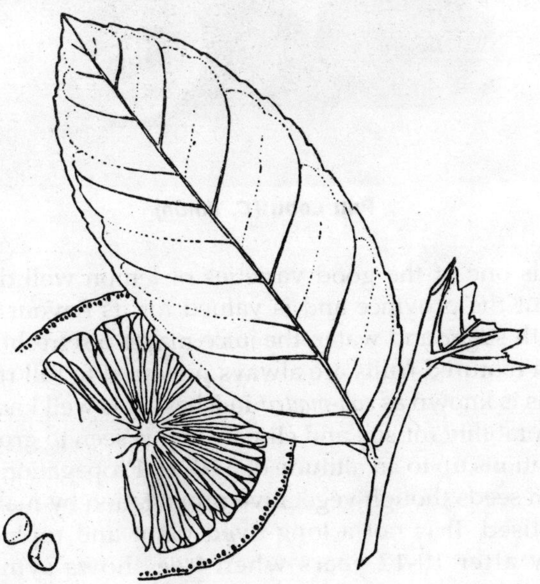

***Pati-lebu (C. limon)*** **showing the outline of leaf and internal structure of fruits**

*Internal fruit characters.* Rind medium thick, thickness 6-8 mm., adherence strong; old glands small to medium, size 0.5-1.5 mm. long, 0.5-1 mm. broad, globose, number 16-20 per ¼ sq. cm., aroma distinct, flavour good mesocarp (albedo) white, sweet, texture hard; axis usually solid, hollow in some fruits; segments 10-13, carpel 7 cm. long, 2.7 cm. broad, 1.5 cm. thick; septa distinct, tough and leathery, dorsal contour slightly convex to truncate; pulp-vesicles white, more or less spindle-shaped, slender, tip acute, size big ones 1.6-2.2 cm. long, 2-3 mm. broad, melting; juice abundant, 3-3½ oz. per fruit, very sour, flavour good; seeds many, 20-100 or more per fruit, shape oval, slightly narrowed towards micropylar end, surface smooth, size 0.9-1.1 cm. long, 5-7 mm. broad, 4 mm. thick, testa cream white, chalazal spot deep purple, tegmen deep brown, cotyledon white, monoembryonic.

**Pati Lebu (*C. limon*)**

This is one of the good varieties of lemon well distributed throughout the province and is valued for its flavour and juice, mixed with sugar and water, the juice makes a refreshing *sherbet*. The green matured fruits are always preferred to full ripe yellow fruits. This is known as *soh-madrit* in Khasi. It is well known for its wide adaptability for soil and climate and is seen to grow even in high elevations up to an altitude of 5,000 ft. Propagation is usually done from seeds though vegetative propagation by marcottage is also practised. It is not a long-lived plant and replacement is necessary after 10-12 years when tree shows symptoms of exhaustion and rapid decline. The variety bears fruit throughout the year with main crop in July to September. It appears to resemble the socalled lemon-citron described under the name *c.medica* var. *bajoura* (Plates CLXII, CLXX, CLXXII) of Bonavia (1890).

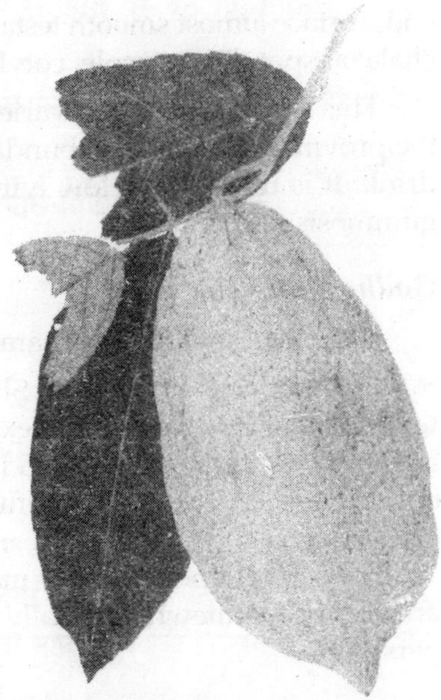

**Jora-tenga (*C. limon*)**

### *Jora-tenga*

Plant, leaf, and flower characters are similar to those of *Pati-lebu.*

*External fruit characters.* Shape oblong to ovate-oblong, gradual tapering towards base and apex; size 12-15.5 cm. ht., 7-8.2 cm. diam.; D/H index 0.53-0.62 (av. 0.55); colour lemon yellow; surface slightly rough or subwarty, coarsely pitted, somewhat ribbed or rugose; base more or less low collared, basal area narrow, 7-9 mm. diam. furrows extending through the collar; stem short. 0.8-1 cm. long; apex mamillate, areole indistinct, somewhat radially ribbed, areolar area protruded into somewhat pointed nipple, stylar scar present, small, usually even, occasionally slightly depressed in a small apical pit, style usually not persistent.

*Internal fruit characters.* Rind medium thick, thickness 6 mm.-1 cm., (usually 6-8 mm.) firmness firm, adherence strong; oil glands conspicuous, number 15-20 per ¼ sq. cm., shape usually oval or globose, size 0.5-2 mm. long, 0.5-1 mm. broad, glandular layer about ⅕ rind; mesocarp (albedo) white, texture hard, sweet; axis semi-hollow or solid, diameter at the centre 1-1.3 cm.; segment 8-12 (usually 10-11), adherence strong, rag medium; septa white, tough, dorsal contour convex to truncate; pulp-vesicle white, texture medium coarse, medium soft. fusiform, size big ones 1.5-2.3 cm. long, 2-3 mm. broad; juice abundant, 3-4 oz. per fruit, very sour; seed many, 15-60 or more, size 9 mm.-1.1 cm. long, 5-7mm. broad, 3-4 mm. thick, shape ovate, slightly attenuated towards micropylar end, surface almost smooth testa cream white, tegmen deep brown, chalazal spot deep purple, cotyledon white, monoembryonic.

This is also a common variety found every where throughout the province. The juice is abundant and makes fine refreshing cold drink. It is unsuited to low lying situations and is susceptible to gummosis and scab.

### *Godha-pati-lebu*

Plant, leaf and flower characters are similar to *pati-lebu*.

*External fruit characters.* Shape ovate-oblong with gradual tapering towards base and apex; size variable, normal 12-13.5 cm. ht., 8.5-9.5 cm. diam.; wt. 1-1.5 lb.; D/H index 0.61-0.75 (av. 0.66); colour lemon yellow, surface pebbled and warty; base low-collared,depressed, basal area rough and rugose; stem short, 1-1.5 cm. long; apex broadly mamillate, areole indistinct, warty, areolar area somewhat radially ribbed,stylar scar even, style not persistent.

**Godha-pati lebu (*C. limon*)**

*Internal fruit characters.* Rind medium thick, thickness 7 mm.-1 cm., adherence strong; oil glands globose to oval, 1-2 mm. long, 1 mm. broad, number 12-14 per ¼ sq. cm., conspicuous; mesocarp (albedo) white, sweet,texture compact and hard; axis usually hollow, diameter at the centre 9 mm.-1.2 cm.; segment (10-13), septa white tough, dorsal contour slightly convex to truncate; pulp-vesicles white, fusiform, tip acute,size 1.2-2 cm. long, 1-2 mm. broad, medium soft; juice abundant, 3-6 oz., intensely sour; rag medium; seeds many, 80-90 or occasionally more per fruit, shape ovate with gradual tapering at the micropylar end, size 1.2 cm. long, 7 mm. broad, 3-4 mm. thick, testa cream white, tegmen deep brown, chalazal cap deep purple, cotyledon white.

The variety is an intermediate form between citron and lemon and may be called as *Citron-lemon.* It is very unattractive owing to the roughness of the skin and is not valued in trade. It flourishes in low lying situations and can withstand certain amount of waterlogging.

***Kata-jamuri***

Kata-jamuri (*C. limon*)

*Plant characters.* A small plant or a shrub, ht. 6-8 ft., densely foliaged; spreading; thorny, usually 5 mm.-1.5 cm. long, rarely 3-4 cm. long; young buds slightly coppery tinged; heavy yielder, season September to November.

*Leaf characters.* Shape oblong; deep green; medium coriaceous; size 8.5-9.5 cm. long, 4.5-5 cm.diam. summer big leaf usually 11-11.5 cm. long, 5.5-6.2 cm. broad, apex almost obtuse to slightly acute, margin crenate; petiole wingless, 7-8 mm. long, articulated.

*Flower characters.* Bi-sexual and staminate, clustered 3-8, cymes; pedicel 4-5 mm. long, peduncle 5-6 mm. long, glabrous; calyx more or less cup-shapped, apex slightly acute or obtuse, purple tinged; corolla imbricate, 4-5 merous, colour light purple outside, oblong, apex slightly acute, size 1.8-2.1 cm. long, 6-8 mm. broad.; stamens 26-40, filaments white, polyadelphous, jointed by 2-3 together and 1 or 2 free, pubescent, 1-1.2 cm. long, anther 4-5 mm. long; ovary usually 4 mm. long, 3 mm. broad; style terete, white, 8 mm.-1 cm. long including stigma, stigma capitate; flowering season mainly in February to March, and also all the year round.

*External fruit characters.* Shape oblong to obovate, gradual tapering towards base, abruptly towards apex; size usually 9.5-10.5 cm. ht., 6.5-6.8 cm. diam. big ones 11.7 cm. ht., 7.5 cm. diam.; wt. 7-12 oz. (av. 8 oz); D/H index 0.65-0.69 (av. 0.66); colour greenish yellow; surface slightly smooth to rugose, puffing, minutely pitted, dull,; base low-necked, occasionally ribbed; stem short, 8 mm.-1.1 cm. long; calyx irregularly divided, thin, tip obtuse; apex abruptly protruded into long and broad nipple, nipple 9 mm. -1.1 cm. long, areole absent, areolar area slightly rough, stylar scar depressed in a small apical pit, style not persistent.

*Internal fruit characters.* Rind medium thick, thickness 5-6 mm., firmness firm, adherence with carpel strong, oil glands small to medium, primary glands globose to oval, 1-2 mm. long, 1 mm. broad, secondary glands 2 mm. long, 2 mm. broad, glandular layer about ⅓ rind gland number 20-24 per ¼ sq. cm., conspicuous; mesocarp (albedo) white, sweet, medium hard; axis usually hollow or semi-hollow, rarely solid, diameter at the centre 1-1.5 cm.; septa chartaceous, tough, white, dorsal contour slightly convex to truncate, carpel size 7-8 cm. long, 2-2.5 cm. broad, 1.5-1.8 cm. thick; segments 8-10, adherence medium strong; rag little;pulp-vesicles usually fusiform, white, size big ones 2 cm.long, 3-4 mm. diam., tip pointed, somewhat coarse, arising from dorsal contour only; juice medium, 3-4 oz. per fruit, very sour, colour white; seeds many 40-60 per fruit, oblong, smooth on both sides, micropylar end somewhat blunt, size 1-1.2 cm. long, 5-6 mm. broad, 4-5 mm. thick, testa white, tegmen light brown, chalazal cap light purple, cotyledon ivory white.

The variety to some extent resembles Eureka lemon in look, but lacks in many of the good qualities with which the latter is endowed. There is a fair amount of juice but it is difficult to squeeze it out due to the vesicles being slightly tough. It is a prolific bearer.

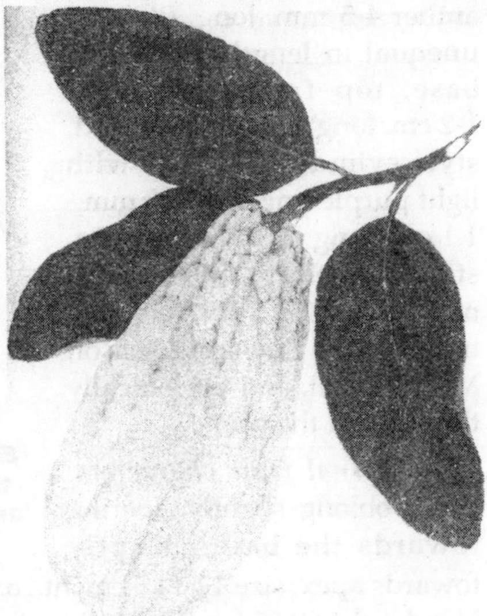

**Elachi-lebu (*C. limon*)**

### *Elachi-lebu (Cardamom-lemon)*

*Plant characters.* A small plant or a shrub, 8-10 ft. high drooping habit, moderately branched, foliage open, thorny, 5 mm. to 3 cm. long, slender; shoots slightly purple; medium fruiter, fruiting season mainly from July to September, very shy crop in other parts of the year.

*Leaf characters.* Leaf medium sized, dark green, coriaceous, size normal ones 10-11.5 cm. long, 5-5.5 cm. wide, big ones 16 cm. long, 8 cm. wide; shape ovate-oblong, apex usually obtuse, margin crenate, slightly revolute, veins slightly prominent on both the surfaces; petiole wingless or rudimentary wing, petiolar wing (if present) 1-3.2 cm. long, 3-5 mm. wide, articulation obscure, slightly articulated on the upper side only, absolutely no articulation mark on the under side, distinct character; crushed leaf aroma distinct, approaching cardamom.

*Flower characters.* Bi-sexual and staminate, proportion of staminate flowers more than bi-sexual ones, glomerate (6-10), cymes; pedicel 5-8 mm. long, peduncle 1.5-2.2 cm. long, glabrous; calyx cupular, greenish white, dotted with oil glands, 4-5, lobed;

corolla imbricate, light purple outside, white inside, petals 4-5 , oblong, size 2.5-3.6 cm. long, 7 mm.-1 cm. broad, dotted with oil glands outside; stamens many, 35-44 or more, anther 4-5 mm. long, filament unequal in length, jointed at base, top free, pubescent, 1-2 cm. long; pistil clubshaped, style cylindrical, white with light purple tinge, style 8 mm.-1.1 cm. long including stigma, stigma capitate, diameter 6 mm.; ovary 1-1.3 cm. long, 5-6 mm. broad; flowering season March-April, and periodically throughout the year.

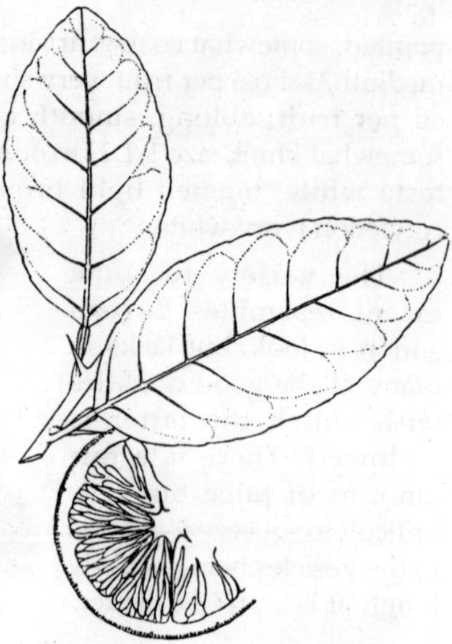

***Elachi-lebu* (*C. limon*) showing the outline of leaf and internal arrangement of fruit characters**

*External fruit characters.* Shape oblong, slightly tapering towards the base abruptly towards apex; size 12-14.5 cm. ht., 6.8-7 cm. diam.; wt. 8-12 oz.; D/H index 0.48-0.65 (av. 0.56); colour lemon yellow; surface sub-warty, slightly smooth towards apex; basal ara rough and rugose, slightly collared; stem 2-3.5 cm long; calyx chartaceous, 4-5 lobed, reflexed upwards, deltoid; apex mamillate, areole indistinct, areolar area somewhat radially ribbed, stylar scar even, style not persistent.

*Internal fruit characters.* Rind medium thick, thickness 5-6 mm., adherence strong; oil glands conspicuous, globose to slightly oval, size small, 1-2 mm. long, 1 mm. wide, number 40-45 per ¼ sq. cm. aroma distinct, approaching cardamom, flavour agreeable; mesocarp (albedo) white, slightly sweet; axis hollow, diameter at the centre 1.2-1.5 cm.; segments 10-13, adherence strong, carpel size 10-12 cm. long, 2.2-5 cm. broad, 1.2cm. thick, septa white, leathery, dorsal contour slightly convex to truncate; pulp-vesicles white, more or less fusiform, irregular in shape, sometimes constricted at two or three places, arising only from dorsal walls, somewhat loosely packed, 1.2-1.6 cm. long, tip pointed, pulp not very melting; juice medium, wt. 3-3½ oz. per fruit, very sour; rag medium; seeds none or sevaeral, 0-16 per fruit, usually seedless, shape almost orbicular to void, micropylar end obtuse, size 7-8 mm. long, 6-7 mm. broad, 5 mm. thick, testa white, tegmen deep brown, chalazal spot deep purple, cotyledon white.

This is generally grown as a novelty in the home-stead gardens. It occurs especially in Karimganj sub-division of Cachar district where the variety is held in great esteem owing to its flavour approaching cardamom, a spice known in vernacular as *elachi (Elettaria cardamomum)* to which it obviously owes its name. It is a shy bearer, fruiting periodically throughout the year and has hardly any commercial possibility.

### *Soh-long*

*Plant characters.* A small tree, 10-12 ft. high; spreading, moderately dense foliage; thorny, usually 1-1.5 cm. long, rarely 3-4.5 cm. long; medium yielder, season September to November, fruit remains in the tree upto January, one crop in a year.

*Leaf characters.* Shape ovate-oblong, coriaceous, yellowish green, size 10-15 cm. long, 7-8.3 cm. broad, apex obtuse, margin crenate, vein prominent on upper side, petiole wingless, usually 1-1.3 cm. long, articulated.

*Flower characters.* Staminate and bi-sexual, proportion of bi-sexual flowers more than the staminate ones, clustered (2-7), cymes; pedicel short, 4-6 mm. long, peduncle 7 mm.-1 cm. long, glabrous; calyx more or less cupshaped, 4-5 lobed, apex somewhat pointed, purple tinged; corolla imbricate, petals 4-5, purple outside, white insie, oblong, size 2.3-2.8 cm. long, 6-9 mm. broad; stamens 27-36, filaments white, glabrous, 0.9-1.6 cm. long; ovary oblong, size usually 5 mm. long, 3 mm. wide, style terete. 1.1-1.2 cm. long including stigma, stigma capitate, 3-4 mm. diam., flowering season March and April.

*External fruit characters.* Shape oblong to obovate; size medium, 11.5-14 cm. high, 8-11 cm. diam. wt. 12 oz.-1½ lb.; D/H index 0.69-0.80 (av. 0.71); colour lemon yellow; surface almost smooth or rugose, dull, section showing through the rind particularly in smooth fruit, bumpy, pitted; base low-collared, slightly depressed; calyx depressed, division irregular, usually tips drying out; stem 1-1.5 cm. long; apex broadly nippled, occasionally slightly nippled, areole indistinct, areolar area radially ribbed, slightly rugose, protruded into both low or prominent nipple, stylar scar present depressed in a small apical pit, style not persistent.

*Internal fruit characters.* Rind medium thick, thickness, 7-9 mm. firmness leathery, adherence medium strong; oil glands conspicuous, number 12-20 per ¼ sq. cm., size 1-3 mm. long, 0.5-2 mm. wide, shape usually oblong with very slightly necked in big glands, glandular layer about ⅓ rind, aroma distinct, similar to one described under the variety *pani-jamir;* mesocarp (albedo) white, spongy and soft, sweet, vascular bundle prominent; axis hollow, diameter at base 0.8-1.5 cm., at the centre 1.6-2.5 cm., at apical end 1.5-1.8 cm.; segments 9-11, adherence medium, septa medium coriaceous, white, somewhat tough, dorsal contour truncate to slightly convex, septa either closed or broken at axis, carpel size 8-10.2 cm. long, 2.2-3.5cm. broad, 1.5-3 cm. thick; pulp-vesicle white, texture coarse, soft and melting, fusiform, tip both blunt and slightly pointed, size big ones usually 1.5-2.5 cm. long 2-4 mm. broad, vesicles arising from dorsal contour; juice plenty wt. 6-8 oz. per fruit, very sour; rag medium; seeds several 6-20 per fruit, oval-shaped, micropylar end usually pointed or breaked, convex on both sides, surface more or less smooth, very minutely anastomosing lines on one side. size 1.2-1.5 cm. long, 7 mm.-1 cm. broad, 5-6 mm. thick, testa white, tegmen light brown, chalazal spot purple, cotyledon white.

It is similar to the *pani-jamir* (sweet lemon) described in the text in all characters except in its acidulous pulp-vesicles which serve to separate the two from each other. This variety is not of common occurrence and has been found to grow only in the Khasi Hills, The fruit has as yet no commercial value though it appears to be suitable for being used in the preparation of lemon squash as it contains fair amount of acid juice. It is a medium fruiter and is highly susceptible to gummosis, scab and canker particularly when grown in low lying situation. Apparently a similar variety by the name *Kumaon* lemon has been pictured in Bonavia's Plate CLXXXVIII (a).

**Soh-synteng (*C. limon*)**

### *Soh-synteng*

*Plant characters.* Plant medium tall, 10-12 ft. high, plants inclined to lodge; moderately dense foliage, spreading; thorny; usually shoots not tinged; fairly heavy fruiter, season early, September and October; one crop in a year.

*Leaf characters.* Leaf medium large, ovate-oblong, pale green, size normal ones 10.5-11.5 cm. long, 6-6.2 cm. broad, summer leaf 14.2 cm. long, 8 cm. braod, apex usually obtuse to slightly acute, margin crenate, veins slightly prominent on both sides; petiole wingless, very rudimentary wing in summer leaf,1.2-1.5 cm. long, articulation obscure.

*Flower characters.* Bi-sexual and staminate, clustered (2-6), occasionally solitary, proportion of staminate flowers less, cymes; pedicel short, 4-5 mm. long, peduncle indistinct, 4 mm. long, glabrous; calyx medium coriaceous, cup-shaped, tips not tinged; corolla imbricate, petals 4-5, light purple outside, oblong, apex more or less acute, size 1.8-3.0 cm. long, 6 mm. -1.2 cm. broad; stamens 24-32, filament white, jointed, almost forming like a tube, very rarely free, glabrous, 1.2-1.6 cm. long, anther 3-5 mm. long; ovary oblong, size usually 4 mm. ht., 3 mm. diam., style terete, greenish white, 1-1.4 cm. long including stigma, stigma capitate, diameter 3 mm.; flowering season March-April.

*External fruit characters.* Shape spherical size 7.4-8.2 cm. ht., 7.5-7.8 cm. diam.; wt. 7-8.5 oz.; D/H index 0.91-1.02 (av.0.97); colour lemon yellow or light yellow; surface smooth and glossy; base evenly rounded; stem short 1.2-1.5 cm. long, not presistent; calyx thin, irregularly divided, tip acute, deltoid; apex slightly nipped to almost obtuse; areole indistinct, areolar area smooth, stylar scar even, style not persistent.

*Internal fruit characters.* Rind thin, thickness 4 mm., leathery, adherence medium strong; oil glands small to medium, primary glands 1-2 mm. long, 1-2 mm. broad, number 14-20 per ¼ sq. cm., globose, conspicuous, aroma distinct, *suigeneris;* mesocarp (albedo) white, slightly sweet; axis hollow, diameter at the centre 1.5-1.7 cm., at apex and base 5-6 mm.; segments 9-12, adherence medium, septa chartaceous, tough, dorsal contour of carpel truncate, carpel size 6-6.8 cm. long, 2.6-3 cm. broad, 1.8-2 cm. thick; pulp-vesicles white, tip obtuse to somewhat acute, fusiform, melting, size big ones 1.8-2 cm. long, arising from dorsal contour, texture medium; juice plenty, 3-5 oz., very sour, flavour good; rag ittle; seeds several, usually 10-18 per fruit, shape oblong, micropylar end round to slightly pointed, size 1-1.1 cm. long, 6-7 mm. broad, 4-5 mm. thick, surface smooth, testa white, tegmen brown, chalazal spot purple, cotyledon slight greenish white.

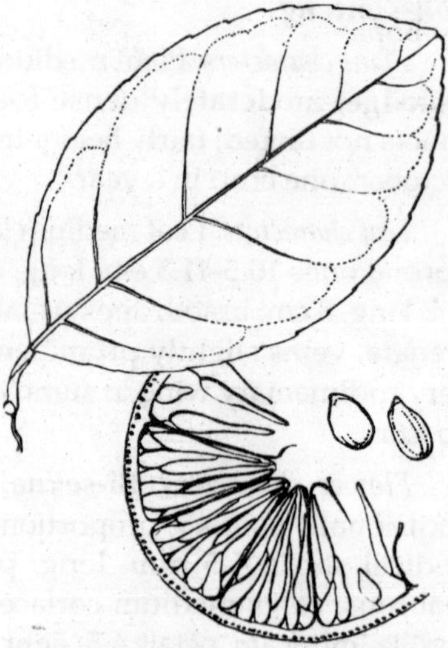

**_Soh-synteng (C. limon)_ showing the shape of leaf and internal structure of fruits**

The variety is usually met with in the submontane tracts. It is known as *soh-kympriak* in the Khasi Hills, *soh-sanikar* in the North Cachar Hills and *sok-synteng* in the Jaintia Hills inhabited by the Synteng tribe from whom the name has probably originated. It is a prolific bearer and yields only one crop in a year. The fruits are very juicy and may profitaby be utilized for bottling lemon squash. The morphological characters of the tree, leaf and the fruit of this variety are so very identical with those of the sweet lime, described in the text elsewhere, that it is impossible to distinguish one from the other unless the fruits are tasted. The *soh-synteng* differs mainly by its purple coloured flower and the chalazal spot and by its intense acid pulp-vesicles as contrary to pure white flower, yellowish white chalazal spot and sweetness of the pulp-vesicle of sweet lime. The *soh-synteng,* it appears is just the counter part of the sweet-lime as found in Assam. It has been tried as a root stock for Valencia Late, Washinton Navel and the Khasi orange and has proved to be incompatible. It is highly susceptible to gummosis and withertip, especially when grown under plain land conditions. An apparently

**Pani-jamir (*C. limon*)**

identical variety under the name *kalan kaghzi* has been referred to by Bonavia in Plate CXCI.

### *Pani-jamir*

*Plant characters.* Tree large, 15-25 ft. high, spreading; thorny, usually 1-4.5 cm. long; moderately dense foliage; slight tendency for lodging; prolific bearer, fruiting season September to December, one crop in a year, plant short lived.

*Leaf characters.* Leaf ovate-oblong pale yellowish green, medium coriaceous, size usually 10-11.6 cm. long, 6.2-7.1 cm. broad, very big ones (summer leaf) 14 cm. long, 8 cm. broad, apex obtuse or slightly acute, margin crenate, very narrowly winged in summer leaf; petiolar wing 1-2.5 cm. long, 4-5 mm. broad, articulated; veins slightly prominent on upper side; aroma assisting diagnosis.

**Pani-jamir (*C. limon*)**

*Flower characters.* Staminate and bi-sexual, proportion of staminate flowers more than the bi-sexual ones, clustered, 2-8, cymes; pedicel 5-6 cm. long, peduncle 7 mm.-2.1 cm. long, glabrous; calyx more or less divided, deltoid, apex tinged, 4-5 lobed; corolla imbricate, petals 4-5, very light purple outside, ephemeral, shape oblong, size 2.4-2.8 cm. long, 0.6-1 cm. broad; stamens 27-34, filaments white, glabrous, polyadelphous, occasionally 1 or 2 free, 1.1-1.7 cm. long, anther 3-5 mm. long; ovary oblong, size 5-6 mm. long, 3-4 mm. broad, style cylindrical, usually white, rarely slightly tinged, 1.1-1.2 cm. long, stigma capitate diameter 3-4 mm.; flowering season March-April.

*External fruit characters.* Shape ovate-oblong; size 13-16 cm. ht., 8.5-11.8 cm. diam.; wt. 1-2 lb.; D/H index 0.65-0.72 (av. 0.67); colour slight greenish yellow to light lemon yellow; surface almost smooth to rugose or occasionally very slightly rough, section showing through the skin, slightly puffing, pitted; base low coloured and slightly depressed, basal area narrow, diameter 0.8-11 cm., rugose, furrows sometimes extending 1-2 cm. through the collar; calyx depressed, division irregular, tips usually drying out; stem 1-1.5 cm long; apex usually prominently and broadly mamillate, occasionally with small-nipple; areole indistinct, areolar area somewhat radially ribbed, style not persistnet.

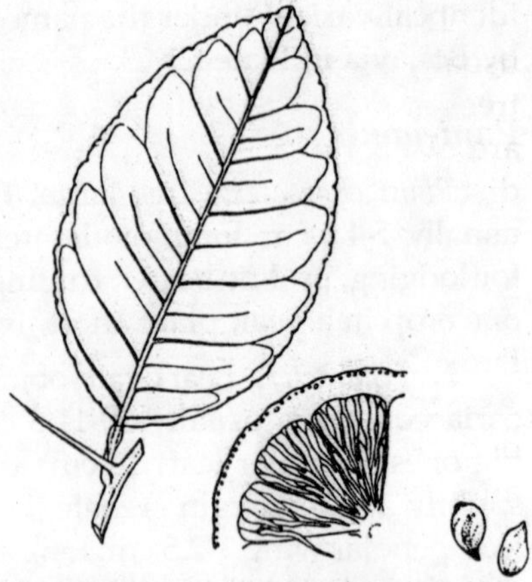

**Pani-jamir (C. limon) showing the outline of lead and internal structure of fruits**

*Internal fruit characters.* Rind medium thick, thickness 6-9 mm., firmness leathery to somewhat soft, adherence medium; oil glands conspicuous, number 12-18 per ¼ sq. cm., size 1-4 mm. long, 0.5-2 mm. wide, oval or ellipsoid, occasionally pyriform, oil abundant, aroma distinct, agreeable; mesocarp (albedo) white, spongy, soft, sweet, vascular bundle prominent; axis hollow, diameter at base 0.7-1.5 cm., at the centre 1-8-2.5 cm., at apical end 1.6-2 cm.; segments 8-11 (usually 10), carpel size 9-11.5 cm. long, 3-3.5 cm. broad, 1.6-3.5 cm. thick, dorsal contour convex to truncate, occasionally retuse, septa-medium coriaceous, white, tough, carpel both closed or broken at axis; pulp-vesicles white, texture coarse, not very melting, shape irregular, usually fusiform, tip usually blunt, size big ones 1.5-2.7 cm. long, 2-4 mm. wide, stalked, arising from dorsal wall; juice abundant, wt. 5-7 oz., sweet or sweetness and acidity blended, flavour good; seeds many, 20-30 per fruit, pointed towards micropylar end, occasionally beaked, surface more or less smooth on both sides, sometimes with minute anastomosing lines, ovate, size 1.3-1.7 cm. long, 7-9 mm. braod, 5-7 mm. thick, testa cream white, tegmen light brown, chalazal spot deep purple, cotyledon white or very light greenish white, polyembryonic.

The variety is rarely seen to grow above 2,000 ft. elevation. It usually features in home gardens as a novelty although sporadic trees are also met with in places where oranges and other citrus are extensively grown. This goes by the name *pani-jamir* in the district of Cachar and Sylhet, *soh-phai* or *soh-pai* in the Khasi Hills and *jutamuri* or *panimuri* or *muri-tenga* in Assam Valley plains. The variety has long been in existence in Assam. Ronbinson (1841) in *Account of Assam* has reported about it and placed it as a sweet lemon by the name *pani-muri-tenga.* The *soh-phai* or *pani-jamir* of Assam has been identified by Tanaka (1937,a) as sweet lime. Both of these (sweet lime and the *pani-jamir*) have almost identical shape and colour of the leaves and this probably had lead him to confusion. It is doubtful if Tanaka had actually seen the flowers and the fruits of the *pani-jamir* or *soh-phai* which has a light purple colured flower (ephemeral), a typical lemon-shaped fruit with a prominent broad nipple and purple coloured chalazal spot as distinct from the sweet lime *(Citrus limettioides* of Tanaka or *C.limetta* of Risso) which has globular fruit of the size of a tennis ball with a light yellow ochre chalazal spot and pure white flower. The *pani-jamir* appears to bear certain amount of resemblance in shape with Dorshapo as described by Webber (1946), though it differs from the latter in some other characters.

The fruits being sweet or rather insipid sweet in taste are liked by many. The flavour is distinct and peculiarly of its own. The juice is plentiful and gives a refreshing cold drink or *sherbet.* As a rootstock it has been tried in Assam and found to be incompatible with the Khasi orange *(C.reticulata* Blanco), Valencia Late, Malta Blood Red, and Washington Navel *(C.sinensis).* The stock portion of the plant invariably shows symptoms of gummosis after a couple of years from the date of budding and this may be attributed to a discordant physiological partnership. The variety is found to be susceptible to scab and canker particularly when grown under low lying plain land conditions.

### The Jambhiri *(Citrus jambhiri* Lushington)

*Origin and name.* The *jambhiri* is probably endemic to India. Many forms of this Citrus occur in different parts of India under different names. This features in the *Memoir of Babar* written in 1519 A.d. Bonavia (1890) mentions that the *jambhiri* is unknown outside India. Hodgson (1937) states that a form of *jambhiri* in India is indistinguishable from the rough lemon of Florida. Swingle (1946) deems it to be widely naturalized or possibly indigenous to British India. Webber (1946) aso mentions about its large scale occurrence in India and suggests that in 1498 or little later it was introduced by Portuguese in India whence it travelled to Europe and other parts of the world. Being a widely employed root-stock, its introduction in different parts of the word was naturally rapid. In Assam it is found to occur in many distinct forms in a semiwild state particularly in the primeval forests.

*Nomenclature.* The *jambhiri* of India, so importnat as a root-stock and so widely known throughout the world as rough lemon is yet in a confused state of taxonomical nomenclature. Reasoner (1888) considered it as an identical hybrid No. 25 of Gallesio (1811). Hume (1941) and Webber (1946) have placed it in the lemon group and described it as a horticultural variety of the lemon, although the latter guesses it to be a probable hybrid of lemon and some type of citron. Swingle (1946) also considers it to be of doubtful origin. Hayes (1945) thinks that the relationship of the true lemon and the *jambhiri* is not very clear and assumes that one might be a hybrid or sport of the other. He is possibly influenced by the recent view of Swingle for in a previous unpublished report he had suggested a specific status for the *jambhiri*. In some of the recent publications the so called rough lemon or the *jambhiri* has been referred to by many authors (Cheema, 1934; Richards, 1938; Naik, 1940; Lal Singh and Sham Singh, 1942; Nandi, Bhattacharya and Dutta, 1943 and 1945; Woodhead and Mouat, 1945) as *Citrus limonia* Osbeck. But Tanaka (1937,a) points out that Osbeck whitle describing *C.limonia* had actually Canton lemon or Rangpur lime in view, which is entirely a different fruit from the true lemon. Thus it is apparent that the correct botanical name of the rough lemon is not *C.limonia* Osbeck as hitherto has been used.

Lushington (1910) was the first to raise the *jambhiri* to the rank of a separate species and named it as *Citrus jambhiri*. Tanaka (1937,a) also recognized it as such, though some of the smooth skinned akin varieties were included by him under *Citrus limonia* Osbeck. Rutherford (1946) also considered *jambhiri* to be distinctly different from the true lemon and called it as *Citrus jambhiri*. Having been convinced of its distinctiveness Richards (1948) designated rough lemon as *C. jambhiri*, although in a previous publication (1938) he placed it under the species *C. limonia* Osbeck. In tree habit, in vigour, in shape of leaf, in external and internal characters of fruits it is so very distinct from the true lemon *(C.limon)* that by no stretch of imagination the two can be classed together under the same species. Besides it is significant to note that the lemons usually flower several times in a year where as the *jambhiri* blooms only once. The *jambhiri* is also distinctive in its resistance to various diseases especially to gummosis to which lemon is usually susceptible. Considering all these basic characters of the *jambhiri* there is probably no justification to debar it from attaining the rank of a species until and unless it has been proved by careful hybridization that the *jambhiri* is of hybrid origin. By keeping the *jambhiri* within the true lemon group (*C.limon* Burm.) it will not only add confusion to the well defined group of lemons but will also militate against the very conception of widely known lemons which are well established in trade and commerce. Besides, different forms, very closely resembling the *jambhiri,* are also in existence to form a well marked group of its own. Such an array of forms of the *jambhiri* are commonly met with in Assam and also in other parts of India. Bonavia (1890) had furnished a number of drawings in his book to illustrate the range of different forms of the *jambhiri* in Plate Nos. CXXX to CXXXVIII. In the light of the foregoing observations the writers favour retaining with advantage the specific name *Citrus jambhiri* Lush. for the rough lemon as accepted by Tanaka and Rutherford.

The varieties of the *jambhiri* occurring in Assam are clearly divisible into two distinct sub-groups on the basis of acidity and sweetness of the pulp-vesicles. The acid fruits which include the well known rough lemon constitute the normal group and are well distributed all over the province to an altitude of 5000 ft. or so. The sweet from is very rare and is usually grown as a novelty in homestead gardens of the plain districts of the province.

## Key To Varieties Of The Jambhiri

A. Acid (pulp) group

B. Tree tall, (ht. 18-25 ft.); spreading; thorny; moderately dense foliage, leaf oblong to lanceolate, light green; petiole wingless, flower solitary, purple coloured, annual bloom.

C. Fruit medium-sized (7-9.5 cm. diam.), oblong to somewhat spherical: areole with deep circular furrow, protruded into broad nipple or pushed up nipple; surface very warty, bumpy, rind greenish yellow, adherence slight; axis hollow;

carpel easily separable; vesicles slight yellowish white; seed clavate to slightly obovate, chalazal cap deep purple.

$V_1$.......*Soh-myndong.*

CC. Fruit globose or oblate; medium-sized, areole with irregular circular furrow, areolar area slightly elevated or even; surface smooth, golden yellow colour; pulp-vesicle white; seeds obovate.

$V_2$......*Kata-jamir.*

CCC. Fruit slight oval to ellipsoid; medium-sized; areolar mark indistinct, protruded into very low nipple; surface smooth; skin bright orange; axis hollow; seeds clavate.

$V_3$......*Soh-jhalia*

BB. Tree medium tall (ht. 18-20 ft.), slightly lodging; moderately dense foliage, leaf deep green, somewhat broad lanceolate; usually wingless; very thorny; flower purple, annual bloom.

C. Fruit almost spherical to slightly oblate; areole distinct, either elevated into disk or slightly depressed, very rarely slightly nippled; surface smooth, skin deep red or scarlet; pulp-vesicle deep orange; axis hollow; seeds clavate.

$V_4$.......*Sinduri-nemu-tenga.*

BBB. Tree small to medium (ht. 10-12 ft.), round top; moderately dense foliage; very thorny; flower purple, annual bloom.

C. Fruit globose, small sized (diameter 4.5-6.5 cm.); surface smooth, colour light lemon yellow; areole distinct with circular furrow, areolar area nearly even or very slightly nippled; vesicles white to yellowish white; seeds small, linear.

$V_5$.......*Nemu-tenga.*

**Soh-myndong (*C. jambhiri*)**

AA. Sweet (pulp) group.

B. Tree medium tall, (18-20 ft.), foliage open; thorny; leaf oblong-lanceolate, dark green, wingless; flower purple coloured, annual bloom.

C. Fruit ovate-oblong, medium-sized; surface warty and bumpy, colour gambose; areole distinct with circular furrow, broadly mamillate; pulp-vesicles colour honey yellow, not very melting; seeds ovate-oblong, chalazal spot very light brown with a flush of yellow tinge.

$V_6$.......*Mitha-tulia.*

## Sour Group

### *Soh-myndong*

*Plant characters.* A medium to large tree, ht. 20-25 ft.; spreading branches; densely foliaged; thorny, thorns usually 5 mm.-4.5 cm. long, stout, sharp; young buds slightly coppery tinged; fruiting season September to December, prolific, one crop in a year.

*Leaf characters.* Leaf medium sized, shape oblong or ob-lanceolate, gradual tapering on both ends, usually light green; petiole wingless or with rudimentary wing, articulated; leaf size, 6-12 cm. long, 4-6 cm. broad, big ones 12-15.5 cm. long, 6-8 cm. broad; petiole 7 mm. -1.5 cm. long; margin (blade) crenate, apex somewhat obtuse, bifid.

*Flower Characters.* Bi-sexual and staminate, proportion of bi-sexual flowers more than the staminate, usually solitary, occasionally clustered (1-4). cymes; pedicel 2-3 mm. long, peduncle 4-6 mm. long, glabrous; calyx light green, 4-5 lobed, more or less cup-shaped; corolla imbricate, petals 4-5, oblong, size 1.6-2.1 cm. long, 5-7 mm. broad, punctate with oil cells on outer side, petals purple outside, white inside; stamens 21-25, filaments white, glabrous, some jointed by 2-3, some free, 8 mm.-1.2 cm. long, anther usually 3 mm. long; ovary 2-2.5 mm. long, 2-2.5 mm. broad, globular, 8-10 loculi, style greenish white, occasionally tinged, 8 mm.-1.1 cm. long, stigma capitate diameter 2-3 mm.; flowering season March to April.

*External fruit characters.* Shape oblong to somewhat spherical; size variable, small ones 6.8-7.8 cm. ht., 7.0-7.5 cm. diam., normal ones 8-9 cm. ht., 7.8-8.5 cm. diam., big ones 9-10 cm. ht., 8.8-9.5 cm.

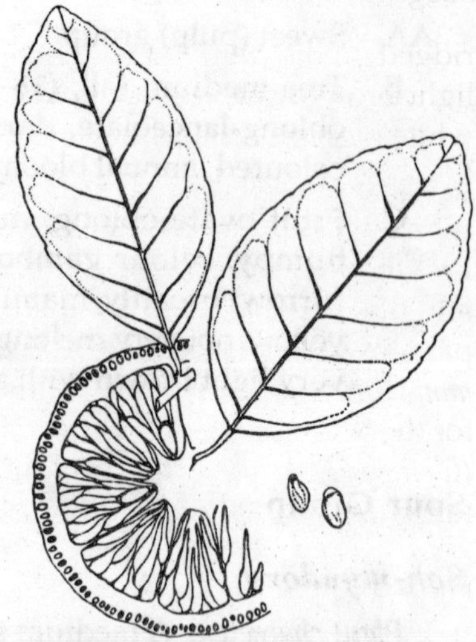

Soh-myndong (C. jambhiri) showing the outline of leaf and internal characters of fruits

diam., wt. 7-11 oz.; D/H index 0.92-1.03 (average 0.97); colour greenish yellow; surface very warty, pitted, bumpy; base short-necked to almost rounded, low-collared, basel area when necked (inside collar) 5-8 mm. diam. surface rugose or rough; calyx small, usually sunken inside the collar, division irregular, 4-5 lobed, tips both pointed or blunt; stem very small, 5-6 mm. long; apex broadly nippled, areole with deep circular furrow or sulcus; areolar area medium broad, diameter 2.5-4 cm., protruded into broad nipple or elevated disk, surface rough and slightly ribbed, stylar scar small, depressed in a small apical pit, style not persistent.

*Internal fruit characters.* Rind medium thick, thickness 6-8 mm., firmness leathery, adherence, slight to medium; oil glands number 19-13 per ¼ sq. cm., conspicuous, primary glands somewhat ellipsoid or pyriform, small ones usually ellipsoid, 1.5-3 mm. long, 1-2 mm. broad, oil abundant, aroma distinct, glandular layer about ½ the rind; mesocarp (albedo) white with very slight yellow tinge towards glandular layer, spongy, sweet, vascular bundle prominent; axis hollow, diameter at the centre 1.5-2.5 cm., at base 8.8-1.1 cm., at apex 1-1.2 cm.; segments 8-10 (usually10), adherence slight; rag little, soft; septa chartaceous, white, tough, dorsal contour convex to truncate, carpel not usually broken at axis, carpel size 5.5-6 cm. long, 2-2.5 cm broad, 1.5-2.5 cm. thick; pulp-vesicles white with very slight yellow tinge, texture somewhat coarse, melting, fusiform, size big ones 1.5-2.2 cm. long, 3-4 mm. broad, usually arising from dorsal side, tip usually blunt, flavour distinct; juicy, wt. 2½-4 oz., colour yellowish white, taste very sour; seeds 20-40 per fruit, rarely more, size usually 1-1.2 cm. long, 6-7 mm. broad, 5-6 mm. thick, shape obovate to slightly elavate, micropylah end slightly beaked or very slightly pointed, surface more or less smooth, very minutely ridged on the rounded end, both sides convex, testa white, tegmen light brown, chalazal spot deep purple or dark brown, cotyledon white or very light greenish white, polyembryonic.

The *soh-myndong* of the Khasi Hills is indistinguishable from the rough lemon of Florida, *jambhiri* or *jamburi* of the Central Provinces, United Provinces and Bombay, *kada narangai* of Madras and the *jathi-khatti* of the Punjab. It is variously known in different parts of Assam such as *godha-tulia* in Assam Valley and *godha-kata-jamir* in Sylhet and Cachar. It has proved to be a compatible stock for the Khasi orange *(C.reticulata)*, the Valencia Late and the *mosambi (C.sinensis)*. It has a wide adaptability for soil and climate and is fairly drought resistant.

### *Kata-jamir*

*Plant characters.* Plant medium tall, ht. 18-22 ft., spreading moderately densely foliaged; thorny, 2-4.5 cm. long., slender; young buds either light green or slightly tinged; fruiting season September to December, one crop in a year.

**Kata-Jamir (*C. jambhiri*)**

*Leaf characters.* Leaf medium coriaceous, somewhat dark green, shape oblong or slightly lanceolate, size 10-11.5 cm. long, 4.5-5 cm. broad, big ones 12-13 cm. long, 5.3-5.6 cm. broad; petiole short 9 mm.-1 cm. long, wingless or rudimentary wing, margin crenate, apex attenuated and obtuse, occasionally emarginate, articulated.

*Flower characters.* Flowers usually solitary or in cluster (1-6), staminate and bi-sexual, cymes; pedicel 3-4 mm. long, peduncle 5-7 mm. long, glabrous; calyx more or less cup-shaped, light green, 4-5 lobed, division distinct, tips acute tinged; corolla imbricate, petals 4-5, purple, size 1.2-1.8 cm. long, 5-8 mm. broad, oblong, slightly reflexed; stamens 24-30, filament white, some jointed by 2-3, 1 or 2 free; glabrous, 8-9 mm. long, anther 2-3 mm. long; style cylindrical, greenish white, 8-9 mm. long, stigma capitate, diameter 2-3 mm.; ovary 3 mm. long, 2 mm. wide, oval, 9-11 loculi; flowering season March to April.

*External fruit characters.* Shape almost spherical to slightly oblate; size 6-7.5 cm. ht., 6.5-7.6 cm. diam.; wt. 6-9 oz.; D/H index 1.00-1.20 (av. 1.06); colour golden yellow, surface smooth, slightly bumpy, minutely pitted; base obtuse, even, finely rugose; calyx slightly sunken 4-5 lobed, deltoid chartaceous; stem mark diam. 4-6 mm., stem short, 5 mm.-1 cm. long; apex obtuse; areole with irregular circular depression, areolar area slightly radially ribbed, slightly elevated, even, diam. of circular area about 2 cm., stylar scar slightly depressed in very small apical pit, scar even and depressed, style not persistent.

*Internal fruit characters.* Rind medium thick, thickness 4-5 mm., firmness leathery, adherence medium; oil glands conspicuous, number 14-16 per ¼ sq. cm., shape primary glands globose to somewhat pyriform, size 2 mm. long, 2-2.5 mm. broad, oil medium, aroma distinct, glandular layer about ¾ of rind; mesocarp (albedo) white, spongy, soft, tasteless; axis hollow, diameter at the centre 1.5-2.5 cm., at base 1-1.5 cm., at apical end 1-1.5 cm.; segments 9-11 (usually 10) adherence slight; rag little, soft; septa chartaceous, white, tough, dorsal contour convex to slightly truncate, carpel size 5.2-6 cm. long, 2-2.6 cm. broad, 1.8-2 cm. thick; pulp-vesicles white or light yellowish white, melting, shape irregular, usually fusiform, size big ones 1.5-1.8 cm. long, 3 mm. braod, arising from dorsal contour, tips attenuated, almost bunt; very juicy, wt. 2.5-4 oz., taste very sour; seeds many, 20-40 per fruit, size normal ones 1-1.2 cm.

long, 5-6 mm. broad, 5-6 mm. thick, shape obovate or clavate, tapering at micropylar end, surface almost smooth, both sides convex, testa white, tegmen light brown chalazal spot deep purple, cotyledon very light greenish white, polyembryonic.

This variety is known as *kata-jamir* in Sylhet and Cachar, *tulia-tenga* in Assam Valley and *soh-jew* in the Khasi Hills. Stephenson and German Jones while giving a list of cultivated varieties of Citrus of the Khasi Hills to Bonavia (1890) mentioned about the *kata-jamir* of Sylhet. Tanaka (1937, a) had recorded this variety under the name C. *limonia* var. *khatta* and pointed out that this variety had been introduced in Alger, and Algeria from India under the name *khatta*. The fruits are smooth and attractive and have a limited demand in local markets. The variety promises to be a good root stock for the Khasi orange *(C.reticulata)*. Valencia Late and Mosambi oranges *(C.sinensis)*.

### *Soh-jhalia*

*Plant characters.* Plant, leaf and flower characters of *soh-jhalia* are similar to those of *soh-myndong*.

*External fruit characters.* Shape usually slightly oval or broadly ellipsoid; size 6.2-8.2 cm. ht., 6-7.5 cm. diam., wt. 4-8 oz.; D/H index 0.89-0.99 (av. 0.93); surface smooth, finely pitted, glossy, sometimes section showing through rind, colour orange yellow, orange colour particularly confined to major half of the fruits towards base; base usually obtuse, occasionally very low nippled; calkx usually even, slightly depressed when necked, division short, tips obtuse, thin; stem small, 5-8 mm. long; apex almost obtuse, occasionally slightly nippled; areole more or less distinct, circular furrows inconspicuous, slightly radially ribbed, stylar scar present, small, depressed in a small apical pit, style not persistent.

*Internal fruit characters.* Rind thickness 3-5 mm., firmness leathery, adherence medium; oil glands few, conspicuous, number 10-13 per ¼ sq. cm. oil glands usually oval to oblate, rarely pyriform, size 1.5-2 mm. long, 1-1.5 mm. broad, occasionally smaller, oil abundant, aroma weak, glandular layer about ½ rind; mesocary (albedo) white to slightly coloured, texture soft, spongy, taste almost tasteless or very slightly sweet, vascular bundle prominent; axis hollow, diameter at base 3-8 mm., at the centre 8 mm.-2 cm., at apical and 5 mm.-1 cm., segments 8-10 (usually 10), adherence medium, carpel size normal ones 5-6 cm. long, 2-2.4 cm. broad, 1.3-2 cm. thick, septa thin white, tough, dorsal contour convex to slightly truncate; rag little; pulp-vesicles light yellow in colour, texture somewhat coarse, melting, usually fusiform, tip attenuated and obtuse, stalked, size big ones 1.2-2.1 cm. long, 2-4 mm. broad., vesicles arising only from dorsal side; juice abundant, wt. 2-4 oz. per fruit, intensely sour; seeds few to many, 10-25 per fruit, size 1-1.4 cm. long, 4-5 mm. broad, 3-4 mm. thick, shape clavate, slender, surface smooth, micropylar end very slightly flat, both sides convex, testa cream coloured, tegmen light brown, chalazal cap deep purple, cotyledon very light greenish white, polyembryonic.

The cultivation of the *soh-jhalia,* the name by which it is known amongst the Khasis, is restricted to the Khasi and Jantia Hills and the North Cachar Hills. It is rarely met with in the plain districts of Assam. The fruit being highly acidic has a restricted demand for domestic use. It is a fast growing tree. It has not yet been tried as a rootstock.

### *Sinduri-nemu-tenga*

***Sinduri-nemu-tenga*** **(*C. jambhiri*)**

*Plant characters.* Plant medium tall, ht., 18-20 ft., spreading, moderately dense foliage; very much thorny, thorns sharp, size 5-6 cm. long, yound buds slightly tinged; fruiting season September to March, prolific fruiter, one crop in a year.

*Leaf characters.* Leaf medium coriaceous, deep green, shape oblong, attenuated gradually on both ends; size 6.3-12 cm. long, 3-6 cm. broad, petiole usually wingless or with rudimentary wing, 6 mm-1.1 cm. long, articulated; blade margin crenate, apex attenuated and slightly pointed, occasionally bifid.

*Flower characters.* Flower solitary or clustered (1-6); bi-sexual and staminate, proportion of bi-sexual flowers more, cymes; pedicel short, 2-3 mm. long, peduncle short, 5-8 mm. long, glabrous; calyx cupular, thin, deltoid, apex tinged; corolla imbricate, petals 4-5, size 1.5-1.9 cm. long, 5-6 mm. broad, shape oblong, apex slightly pointed, colour purple outside, white inside; stamens 22-28, filament white, some jointed by 2-3 and some free, glabrous, 9 mm.-1.3 cm. long, anther 2.5-3 mm. long; ovary globular, 2.5 mm. long, 2-2.5 mm. broad, style slender, greenish white, cylindrical, 1-1.1 cm. long, stigma capitate, diameter 2 mm.; flowering season March to April.

*External fruit characters.* Shape usually globose to oblate, size 4-7 cm. ht., 4.5-7.5 cm. diam.; wt. 2.5-7 oz. (normal weight 4-5 oz.); D/H index 0.97-1.08 (av. 1.02); colour deep or searlet; surface smooth, section showing through rind, somewhat puffy, finely pitted, not very glossy; base obtuse, occasionally very slightly necked, if necked usually ribbed or furrowed; stem short, 6-8 mm. long, persistent, fruits never drop off naturally; apex slightly depressed or almost rounded, occasionally slightly nippled; areole distinct with circular furrow, diameter 1.5-2.5 cm., very slightly nippled, stylar scar present, depressed in a small apical pit, style not persistent.

*Internal fruit characters.* Rind thickness 2-4 mm., firmness soft, adherence medium; oil glands conspicuous, number 15-18 per ¼ sq. cm., shape usually oblate to oval, size 1-1.5 mm. ht., 1-2 mm. diam., aroma distinctly like that of the *jambhiri* glandular layer thick, more than ½ rind; mesocarp (albedo) thin, slightly orange tinged, spongy, sweet; axis hollow, diameter at base 7 mm.-1 cm., at the centre 1.5-2 cm., at apical end 8 mm.-1.3 cm., vascular bundle prominent; segments 8-11, (usualy 9), adherence slight, carpel easily separable, carpel size 4.5-6 cm. long, 1.5-2.2 cm. broad, 1.7-2.5 cm.

thick, septs very thin, slightly tough, dorsal contour convex to slightly truncate; rag little, soft; pulp-vesicles coarse, orange colour, melting, shape irregular, usually fusiform, tips blunt, stalked, arising only from dorsal contour; juice plenty, 1-3 oz. per fruit, intensely sour; seeds few to many, 2-22 per fruit, (usually 15-20), small, size 1.1-1.2 cm. long, 5-6 mm. broad, 4-5 mm. thick, shape clavate to oblong surface almost smooth, minute anastomising lines on both sides, testa cream white, tegmen light brown, chalazal spot deep purple, cotyledon light greenish white, polyembryonic.

The *sinduri-nemu-tenga* of Assam compares favourably with the so called Rangpur lime as described by Webber (1946). *Sinduri* which in English means vermilion, is an appropriate prefix to the local name for denoting the characteristic vermilion colour of the fruit. Bonavia (1890, Plate CXIV) considered Rangpur lime as similar to *suntara* or sour mandarin. Tanaka (1937, a) found no difference between Rangpur lime and *surkh-nimboo* of Lucknow and considered these to be a form of *C.limonia* Osbeck. Rutherford (1946) followed Tanaka. Swingle (1946) suspected Rangpur lime to be a dilute backcross hybrid of *C.reticulata.* var. *austera*. Matlack (1931) as quoted by Webber (1946) suspected Rangpur lime to be a hybrid of lime and mandarin or a sour mandarin orange. Webber, however, provisionally places Rangpur lime under lime group.

The *sinduri-nemu-tenga* possesses some typical characters of sour mandarin with regard to shape and size of leaf, red colour of fruits and the pulp-vesicles, greenish cotyledon and it also possesses the peculiar areolar mark, aroma of rind, deep purple coloured chalazal spot and size and shape of fruit of the *jambhiri* group. It has, however, nothing in common with the lime. Even if it be considered as a hybrid it is probably one of *C.reticulata* var. *austera, i.e.,* sour mandarin and some form of the *jambhiri* or even lemon but certainly not of lime. As *sinduri-nemu-tenga* conforms to most of the morphological characters of the *jambhiri* except in the colour of the rind and the vesicles, it has provisionally been placed as a form of *jambhiri* by the authors. This variety has not yet been tried as a root-stock in Assam but it is not likely to be a serviceable stock as it is very much susceptible to gummosis and wither-tip under Assam conditions. The fruit is seldom eaten but is often used for decoration during festivals. The variety is mostly confined to places of Assam Valley close to Rangpur, a district in Bengal from where, it is believed by many to have originated. It is seldom met with in the hills and is considered to be an introduced variety in Assam.

## *Nemu-tenga*

**Nemu-tenga (*C. jambhiri*)**

*Plant characters.* A small tree, ht. 10-12 ft.; spreading; rounded top; foliage medium open; very slight tendency for lodging; very thorny, 1-6 cm. long, young shoots slightly tinged; plants supposed to be short-lived; fruiting season September to December, one crop annually.

*Leaf characters.* Leaf small, medium coriaceous, light green, shape oblong to slightly ob-lanceolate, articulated; petiole very short, wingless, 5 mm.-1.1 cm. long, blade small ones 5.4-7 cm. long, 2.7-3.4 cm. broad, normal ones 9-10.5 cm. long, 4.2-5.3 cm. broad, big summer leaves 11.5-13 cm. long, 5.5-6 cm. broad, margin crenate or bi-crenate, apex somewhat acute emarginate.

*Flower characters.* Usually solitary or in pairs, rarely in clusters of 3-4, cymes; peduncle 3-4 mm. long, pedicel 2-3 mm. long, glabrous; calyx light green with purple tip, 4-5 lobed; petals purple outside, white inside, shape oblong, size 1.2-1-6 cm. long, 4-6 mm. broad: stamens 21-25, filament glabrous, white, some free and some jointed at the base by 2-4 together, top free, 8 mm.-1.1 cm. long, anther 2-3 mm. long; ovary usually 2 mm. ht., 2 mm. diam., somewhat globular, 8-10 loculi, style greenish white, cylindrical, 8-9 mm. long, stigma capitate, 2 mm. diam.; flowering season March to April.

*External fruit characters.* Shape spherical or globose; size variable, small ones 4.2-4.8 cm. ht., 4.4-5 cm. diam., normal ones 5-5.5 cm ht., 5.1-5.6 cm. diam.; wt. 2-5 oz. (av. normal weight 3 oz.); D/H index 1.00-1.09 (av. 1.03); colour light yellow; surface smooth, section showing through rind, minutely pitted; base round and even; calyx thin, small, irregularly divided, deltoid, tip blunt; stem very short, 5-7 mm. long; apex obtuse, areole an irregular circular depression, usually half of the circular mark distinct, giving an appearance of crecent, diameter 1.2-1.5 cm., usually even, surface almost smooth to slightly rugose and radially ribbed, stylar scar small, depressed in a very small apical pit, style not persistent.

*Internal fruit characters.* Rind thin, thickness 2-3 mm., firmness leathery; adherence medium; oil glands prominent, number 18-22 per ¼ sq. cm., shape globose, size 1.5 mm. hit., 1.5 mm., diam.; oil little, aroma distinct, peculiar, glandular layer slightly more than ½ rind; mesocarp (albedo) white, spongy, tasteless; axis semi-hollow or hollow, diameter at the centre 5 mm.-1.2 cm., at base 3-8 mm., at apical end 4-9 mm.; segments 8-10 (usually 10), adherence medium; rag little, soft; septa very thin, white, somewhat tough, dorsal contour convex to sightly truncate, carpel size 4.0-4.5 cm. long, 1.5-2 cm. broad, 1.2-1.6 cm. thick; pulp-vesicles usually white or light yellowish white in too ripe fruits, melting, shape irregular, fusiform, big ones usually 1.1-3 cm. long, 2. mm. broad, arising from dorsal wall of the carpel, flavour good; juicy, one ounce or slightly more, very sour; seeds few, 6-12 per fruit occasionally slightly more, size 1-1.1 cm. long, 4-5 mm. broad, 3-4 mm. thick, shape clavate to linear, somewhat pointed towards micropylar side, surface almost smooth, both sides convex, testa white, tegmen light brown, chalazal spot deep purple, cotyledon white to light greenish white, polyembryonic.

This variety is very popular with the Assamese and it usually meets the demand for true sour lime. Often it is erroneously called a lime (*kaghzi*) because of its smallness of size and some resemblance with round sour lime. Robinson (1841) in his book *Account of Assam* mentioned this variety apparently wrongly under acid lime. The *nemu-tenga* of Assam has been illustrated by Bonavia (1890, Plate No. CCXXXIX).

## Sweet Group

### *Mitha-tulia*

*Plant characters.* A medium tall plant, ht. 18-20 ft., spreading; moderately dense foliage; very thorny, thorn big ones usually 3-6 cm. long; young buds light green; fruiting season October to February, one crop annually, prolific fruiter.

*Leaf characters.* Leaf medium coriaceous, wingless; shape somewhat lanceolate to ob-lanceolate; size small ones 7-9 cm. long, 3-4 cm. broad with petiole 1-1.2 cm. long, big ones 13-13.6 cm. long, 5.5-7 cm. broad with petiole 1.1-1.5 cm. long, leaf somewhat abruptly attenuated at the apex; deep green, articulated, margin crenate, apex almost caudate, acute, emarginate; veins prominent on lower side.

*Flower characters.* Staminate and bisexual, solitary and clustered (1-8); pedicel 3-4 mm. long, peduncle indistinct. 5-6 mm. long, cymes, glabrous; calyx light green, tips ciliate, slight purple tinged, 4-5 lobed; corolla imbricate, petal 4-5, oblong, size 1.6-2.2 cm. long, 7-8 mm. broad, colour purple outside, apex somewhat acute; stamens, 24-27, filaments white, polyadelphous, usually top free, glabrous, 0.9-1.4 cm. long, anther 3-4 mm. long; ovary ht. 2-3 mm., 2-3 mm. diam., style 0.9-1.2 cm. long including stigma, terete, greenish white, stigma capitate, 2 mm. diam.

*External fruit characters.* Shape oblong or ovate-oblong; size 6-7.8 cm. ht., 6.0-7 cm. diam., big ones 8.5-9.2 cm. ht., 6.8-7 cm. diam.; wt. 3-7.5 oz. (normal wt. 4-6 oz.); D/H index 0.80-0.95 (av. 0.86); colour greenish yellow; surface very warty, usually bumpy, pitted, dull; base usually short necked or low collared, depressed, basal area (inside collar) 5-8 mm. diam., more or less ribbed, furrows extending through the collar upto 2 cm., usually 1 cm. or slightly less; calyx depressed, small, division irregular, thin, tips dried up; stem short, 5-7 mm. long; apex with a prominently circular furrow, areolar area medium, diameter 1.6-2.5 cm., protruded into broad nipple or pushed up nipple, surface wrinkled, slightly radially ribbed, style not persistent.

*Internal fruit characters.* Rind thin to medim, thickness 5-8 mm., firmness leathery, adherence medium strong, oil gland number 25-32 per ¼ sq. cm., conspicuous, shape usually oval to ellipsoid, sometimes pyriform, size usually 1-2 mm. long, 1-1.5 mm. wide., oil medium, aroma weak, glandular layer slightly more than half rind; mesocarp (albedo) cream white, spongy, taste sweet, vascular bundle prominent; axis hollow, diameter at base 5 mm.-1. cm., at the centre 1-1.5 cm., at apical end 8 mm.-1 cm., segment 8-10 (usually 10), adherence medium to slight, septa medium coriaceous, tough, white, dorsal contour convex to truncate; carpel size 5-6 cm. long, 1.8-2.1 cm. broad, 1.3-2 cm. thick; pulp-vesicles very light yellow or honey yellow, texture coarse, not very melting, shape irregular, usually fusiform, size big ones 1.2-1.6 cm. long, 2-3 mm. broad, tips attenuated and blunt, stalked, 8 mm. long, stalk yellow coloured, arising from the dorsal side fo the carpel, somewhat loosely packed; juice scanty, 0.5-1.5 oz., colour light yellow, sweet; seeds many 15-30 per fruit, size 9 mm. -1.2 cm long, 5-6 mm. board, 4-5 mm. thick, shape oblong to ovate, smooth, convex on both sides, micropylar end somewhat blunt, testa cream white, tegmen light brown, chalazl cap light brown with a fiush of yellow tinge, cotyledon very slightly greenish white or white.

The *mitha-tulia* is an acidless form having no other recognizable morphological difference with the true *jambhiri.* It is found sporadically in the plains and seldom, if ever, in the hills of Assam. There is no commercial importance of this fruit as it is neither very juicy nor very tasteful, though sweet. A reference of a similar acidless form of *jambhiri* in the Punjab has been made by Tanaka (1937, a). It is susceptible to gummosis.

## The Karna (*Citrus karna* Rafinesque)

*Origin and name. Citrus karna* is represented by *soh-sarkar* or *soh arsarakar* of the Khasi Hills. The word *soh-sarkar* indicates that it is an introduced fruit in Assam by the Government, as *soh* stands for fruit and *sarkar* for the Government in the Khasi language. So far ascertained, it was first introduced by late B.C. Basu, formerly Deputy Director of Agriculture and Land Records of the East Bengal and Assam, sometimes in the years 1904-1906 when he started a tropical fruit plantation in Wahjain near Cherrapunji (Khasi Hills). This variety is not met with anywhere else. Lal Singh and Sham Singh (1942) pointed out that the *soh-sarkar* of Assam is synonymous with the *kharna-khatta* of the Punjab., *id-lemon* of Poona and Bombay and *mokari* of Renala Khurd (Punjab). Bonavia (1890) dealt with *karna* and *khatta* exhaustively in his *Oranges and Lemons of India and Ceylon.* The *karna* is probably indigenous to India.

*Nomenclature.* In the current horticultural literature of India *kharna-khatta, id-nimbu* and *karna* are frequently mentioned. These have been identified as *C. karna* Raf. by Lal Singh and Sham Singh (1942), while Bonavia (1890) proposed to call it *C.aurantium* var. *khatta,* obviously he had confused this with the seville or sour orange. Swingle (1946) listed *C.karna* Raf. as synonym for *C.aurantium* Linn. Gandhi (1934) opines that true *khatta* is indistinguishable from the true sour orange of commerce and the *karna* of India is distinctly different from the sour orange as well as from the rough lemon *(jambhiri)* although there is a certain amount of similarity in its plant and leaf characters with those of the rough lemon. From his study of the fruit it is evident that the *karna* of India cannot be classed with sour orange *(C.aurantium)* as is done by many. Hayes (1945) points out that *karna* or *khatta-karna* has been suspected to have originated by natural crossing between the sour orange and the lemon and the fruit is designated by him as *C. karna* Raf. Tanaka (1937, a) identifies *khatta, karna,* or *id-limbu* of Bombay, United Provinces and the Punjab, as belonging to *C.karna* Raf. This has been renamed by Lushington (1910) as *C.dimorphocarpa.*

The *soh-sarkar* of Assam and the *kharna-khatta* of the Punjab have been studied side by side at the Citrus Fruit Research Station, Burnihat and both have been found to be identical having no morphological difference whatsoever between them. Detailed studies also reveal that it is distinctly different from either of the well known rough lemon and the sour orange and it will be unwise to include *C.karna* with them. It is apparent that *karna* deserves a separate specific status till it has been proved to be definitely a hybrid or otherwise. Besides, there is a long standing usage behind it in India as to justify such a specific recognition. The *soh-sarkar* of Assam has, therefore, been treated here as a variety under species *C.karna* Raf., the oldest valid name for this class of *Citrus* fruits.

## Key to Variety of the Karna

A. A medium tall tree; spreading; thorny; leaf ovate, petiole wingless or very narrowly winged; flowers both bi-sexual and staminate, purple, annual bloom.

B. Fruits variable, large, usually ovate-oblong, broadly mamillate, occasionally almost obtuse, size 9-12 cm. ht., 8-11 cm. diam.; rind thickness 0.5-1.8 cm., mesocarp usually white, spongy, very sweet; vesicles orange coloured, melting, very juicy, sour; seeds many, chalazal spot purple, cotyledon white.

V...........*Soh-sarkar*

### *Soh-sarkar*

*Plant characters.* A medium tall tree, ht. 15-20 ft.; well branched, drooping very much when loaded with fruits; slight tendency for lodging, moderately dense foliage with open centre; thorny; young shoots tinged; fruiting season December to February, prolific fruiter, one crop annually.

*Leaf characters.* Leaves medium-sized, petiole wingless or very narrowly winged; ovate-oblong, medium coriaceous, dark green, articulated; size 6.5-9.5 cm. long, 4.5-5.5 cm. broad with petiolar wing 0.8-1.6 cm long, 2-3 mm. broad, big ones 11-11.5 cm. long, 7-7.2 cm. broad, margin usually serrate, very rarely crenate, apex attenuated and obtuse, occasionally emarginate.

*Flower characters.* Bi-sexual and staminate, proportion of staminate flowers less than bi-sexual ones, cymes solitary and clustered, (1-6); pedicel 5-7 mm. long, peduncle 5-6 mm. long, glabrous; calyx medium coriaceous, more or less cupular, tips slightly acute or obtuse, purple tinged, 4-5 lobed; corolla imbricate, petals 4-5, colour purple outside, white inside, shape oblong, tips attenuated and somewhat obtuse, size 2.2-2.7 cm. long, 5-8 mm. broad; stamens 24-29, filaments jointed at the base, top free, white, glabrous, 9 mm.-1.8 cm. long, anther 3-4 mm. long,; ovary ovate, 4 mm. ht., 3 mm. diam., style terete, greenish white, 1.2-1.4 cm. long, stigma capitate, 3 mm. diam.; flowering season March-April.

*External fruit characters.* Shape variable, usually ovate-oblong, occasionally almost spherical; surface almost smooth to slightly undulating, finely pitted; colour golden yellow to light orange; size 7-12 cm. ht., 6.5-11.2 cm. diam., wt. 10-20 oz., usually 12-15 oz.; D/H index 0.80-1.02 (av. 0.91); base obtuse in small fruits, others slightly truncate or slightly depressed, basal area slightly depressed, ribbed; calyx coriaceous, irregularly divided, tips either acute or obtuse; stem short, 8 mm-1 cm. long; apex usually very broadly mamillate, occasionally obtuse, stylar scar present, small, depressed in a small apical pit, style not persistent.

**Soh-sarkar (*C. Karna*)**

*Internal fruit characters.* Rind thick, thickness 0.5-1.8 cm. (usually 1-1.5 cm. thick), adherence medium to strong; oil gland medium large, size 0.5-3 mm. ht., 0.5-1.5 mm. diam., usually ovate or elliptical, small ones globose, big ones pyriform, conspicuous, number 20-35, per ¼ sq. cm., oil medium, aroma mild, distinct, glandular layer about 1/5 rind, mesocarp (albedo) white, soft, spongy, taste very sweet; axis solid, spongy, diameter at base 8 mm. -1.1 cm., at the centre 7 mm.-1.6 cm., at apical end 7 mm.-1.5 cm.;

**_Soh-sarkar (C. karna)_ showing leaf character and internal fruit structure.**

segments 9-13, (usually 10-11), adherence medium, carpel size 5.5-7.2 cm. long, 2-3.1 cm. broad, 1.2-2 cm. thick, septa white, slightly tough, medium coriaceous, dorsal contour usually truncate or slightly concave to slightly wavy; pulp-vesicles orange colour, texture coarse, shape irregular, fusiform, tips both acute or obtuse, melting, size big ones 1-2.2 cm. long, 2-4 mm. broad, stalked, vesicles arising from the dorsal side and also from the lower end of the radial walls; juice plenty, 2-4 oz., colour light orange; rag soft, medium; seeds many 20-30 per fruit, size 9 mm. -1.5 cm. long, 6-7 mm.

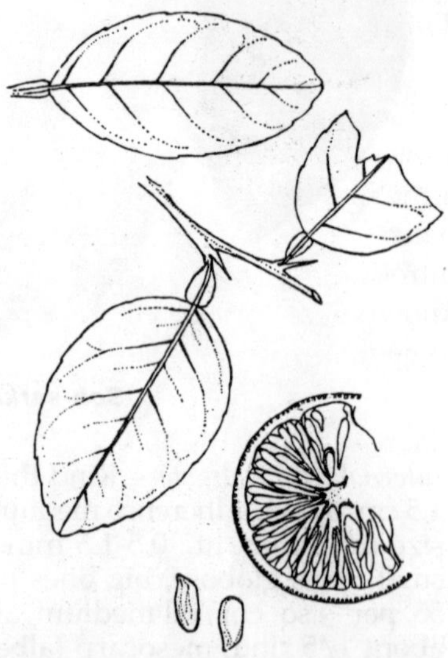

**_Abhayapuri lime (C. aurantifolia)_ showing the outline of leaves and internal structure of fruit**

broad, 4-6 mm. thick, oblong, occasionally somewhat globose, micropylar end flat in oblong ones, almost round in spherical ones, surface one side plano-convex, other marked by anastomosing lines, in some seeds anastomosing lines on both sides, testa white to light cream white, tegmen brown, chalazal cap purple, cotyledon white.

As stated earlier the cultivation of the *soh-sarkar* is confined to the southern slopes of the Khasi Hills where it is grown as a novelty. The fruits being very sour and worthless for any commercial use at present, have not attracted any attention. It has not been found to be serviceable as a root-stock for the Khasi orange (*C. reticulata*), in Assam. The plant on its own root suffers very badly from gummosis under Assam conditions.

**Karimganj lime (*C. aurantifolia*)**

## The Sour Limes (*Citrus aurantifolia* (Christm.) Swingle)

*Origin and name.* Lime is considered almost definitely to be indigenous to India (Royle, 1839, Wight, 1840-50, de Candolle, 1886/1912, Hooker 1872, Brandis, 1906; Tolkowsky, 1938), although Swingle (1946) holds the view that it is apparently a native of the East Indian Archipelago. Most probably lime in an introduced fruit in Assam but when and how it become feral is not known. It is very much well known and popular with the indians in general and is prized for its various uses in every day life.

Real sour lime is known as *kaghzi* almost everywhere in India. The name *kaghzi* probably owes its origin to the Indian word *kagaj* the paper, with which the rind of the fruit is always compared for its thinness. The word *kaghzi* is variously spelled as *kaghzi* (Bonavia, 1890), *kaghuzee* (Burns, 1938), *kagdi* (Cheema and Dubhashi, 1938), and *kagzi* (Javaraya, 1934). But all these words mean one and the same thing. The word *kaghzi* is usually accepted now in most of the current literature where real sour time is meant.

*Nomenclature.* In old taxonomic literatures the limes and the lemons are very much confused and usually one stands for the other. This probably is due to the fact that the distinction between the two was imperfectly understood and that the early writers considered the limes, the lemons and the citrons to belong to one and the same species *C. medica* Linn. Roxburgh (1832) called limes under a separate species *C. acida* which had also been accepted by Lushington (1910). But the lime had been placed as a variety under the citron by Hooker (1872), Brandis (1906). The former named it as *C. medica* var. *acida*. Most of the subsequent Indian writers followed Hooker's nomenclature. Swingle (1913, a) clearly established the position of the lime as an independent species and called it as *C. aurantifolia*. (Christm.) Swingle. This nomenclature has now universally been accepted for sour limes. For different scientific names of lime, offered by different taxonomists from time to time, a reference may be made to the *Citrus Industry* Vol. I., Pp. 401.

**Karimganj lime (*C. aurantifolia*) showing the outline of leaf and internal structure of fruit**

The real taxonomic position of the so called sweet lime is very much anomalous. Webber (1946) gave sweet lime a varietal status under the sour lime while dealing with the cultivated citrus varieties of America. The present authors hold a different view and have placed it under a separate species *C. limetta*. The justification of separating the sweet lime from the sour lime (*C. aurantifolia*) has been fully discussed while dealing with *C. limetta*.

There are only two distinct varieties of the sour lime very commonly met with in Assam. These two varieties have been keyed below.

## Key to Varieties of the Sour Lime

A. Small tree or shrub, bushy habit; spinous, slender; leaf small, elliptical to oblong, petiolar wing small but prominent, leaf aroma distinct, typical of lime; flowers all the year round, petals white.

B. Fruit very small like pegion's egg, diameter 4-5 cm. (1½-2 in.); surface smooth and glossy, colour greenish yellow; very slightly nippled or pushed up nipple; pulp-vesicles greenish white, very sour; seeds few, small, chalazal spot dark brown, cotyledon greenish white.

C. Shape elliptical or egg-shaped $V_1$.....*Abhayapuri-lime.*

CC. Shape spherical or round $V_2$.......*Karimganj-lime*

***Abhayapuri-lime.***

*Plant characters.* A small tree or a shrub, full grown tree 6-10 ft. bushy habit, moderately dense foliage, branch slender, drooping; thorny, small, sharp, 0.6 cm.-0.5 cm. long, slender; prolific bearer, peak season August to October, minor crops all the year round.

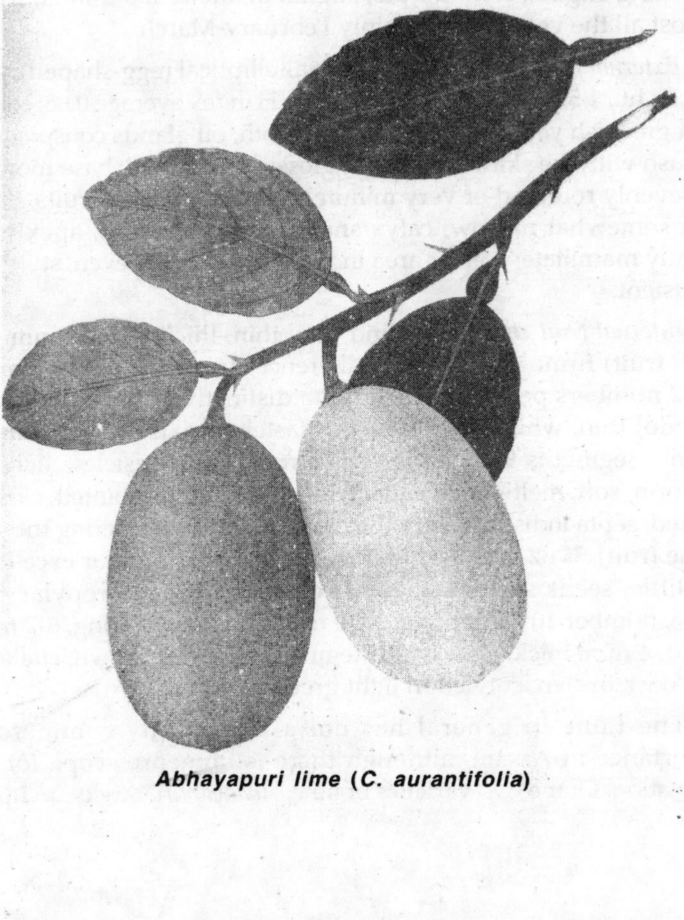

**Abhayapuri lime (*C. aurantifolia*)**

*Leaf characters.* Leaf small with prominent wing, blade elliptical to oblong, ovate, size 6-7.5 cm. long, 3.5-5 cm. broad, apex obtuse, margin crenate; petiole winged, shape spathulate to obcordate, size 1.2-1.5 cm. long, 0.1-13 cm. broad, apex obtuse to slightly retuse, margin crenate or entire, leaf dark green, chartaceous, articulated.

*Flower characters.* Flowers small, usually glomerate (2-8), lax racemes; pedicles 3-4 mm. long, peduncle 7-9 mm. long, glabrous; calyx small, 4-5 lobed, apex acute, deltoid, greenish white dotted with oil cells; corolla white both inner and outer side, (in hot climate very slightly tinged at bud stage but ephemeral), petals 4-5, size 1.2-1.6 cm. long, 4-6 mm. broad, oblong, dotted with oil glands on outside; stamens 22-26, anther 2-3 mm. long, filaments white, jointed 3-4 at the base, top free, 6-9 mm. long; ovary 4-5 mm. ht., 3 mm. diam., ovalshaped, style cylindrical, short 3-4 mm. long including stigma, style purple, stigma diameter 2-3 mm., flowers almost all the year round, mainly February-March.

*External fruit characters.* Form oval-elliptical (egg-shaped); size 6-7 cm. ht., 4-5 cm. diam., wt. 2-3 oz.; D/H index average 0.68; colour light greenish yellow, surface very smooth, oil glands conspicuous in flush with the skin, also pitted, glossy; stem small; base more or less evenly rounded or very minutely necked in some fruits, basal area somewhat narrow; calyx small, thin, tips acute, apex very slightly mamillate; areolar area indistinct, pistil scar even, style not persistent.

*Internal fruit characters.* Rind very thin, thickness 1-3 mm. (in fresh fruit) firmness leathery, adherence strong, oil glands small, 28-32 numbers per ¼ sq. cm., aroma distinctly of lime; mesocarp (albedo) thin, white, sight sweet or tasteless; axis solid, diameter 5 mm.: segmeuts 9-10, adherence strong; pulp-vesicles, slender, fusiform, soft, melting, greenish white, tip blunt or pointed, closely packed, septa indistinct, very thin; juice plenty (considering the size of the fruit), ¾ oz. to 1.5 oz. per fruit, very sour, flavour excellent; rag little; seeds small, somewhat ovate-oblong, micropylar end acute, number 10-14 per fruit, size usually 1-1.1 cm. long, 6-7 mm. broad, 4 mm. thick, testa white, tegmen very light brown, chalazal spot dark brown, cotyledon light greenish white.

The Lime in general has not assumed any commercial importance in Assam, although there is immense scope for its cultivation. Of the two varieties of lime, *Abhayapuri* variety occupies a considerable area and is particularly found to grow in Pathsala, Pathasarkuchi, Bongaigaon and Abhayapuri in the district of Goalpara. This variety mostly predominates in the Assam Valley, although round type of lime is not altogether absent. Most of the production finds its way into Calcutta and Gauhati markets where demand for lime is very great. Being prolific and having comparatively bigger fruits this variety is particularly recommended for large scale production. It is extensively used for various culinary purposes and for preparing pickles, *chutneys* and *sherbet*.

The variety is very much susceptible to gummosis, scab and canker and as such control measures are always essential for successful cultivation. Lime is supposed to be short lived and the yield decreases progressively with advancing age. Regular cultural treatments, especially manuring, are necessary to keep up the high level of production. Propagation is generally carried out by marcottage or air layering and occasionally by seeds.

### *Karimganj-lime*

*Plant characters.* A small tree or a shrub, bushy habit, ht. 6-10 ft. spreading, branch slender, densely foliaged; thorny, short, sharp, 0.5-1.5 cm. long; heavy yielder, peak season August to October, periodical shy crops almost all the year round.

*Leaf characters.* Leaf small and prominently winged; lamina elliptic oval or ovate-oblong, deep green, size normal 6-7.5 cm. long, 3.8-4.6 cm. broad, apex obtuse, margin crenate; petiolar wing broadly spathulate or cordate, size normal 1.2-1.5 cm. long, 5-7 mm. broad, apex slightly truncate or obtuse, articulated, margin crenate, young bud light green or very light coppery tinged.

*Flower characters.* Flowers small, usually glomerate, (2-8), staminate and bisexual, recemes; pedicel 3-4 mm. long, peduncle 5-7 mm. long; calyx small, 4-5 lobed, deltoid, apex acute, colour greenish white, dotted with oil glands; corolla imbricate, white, very slightly tinged at bud stage, ephemeral, petals 4-5, size 1.0-1.5 cm. long, 4-6 mm. broad, oblong, dotted with oil gland outside, apex obtuse; stamens 20-28, anthers, 3-5 mm. long, filament white, glabrous, jointed by 3-4, polyadelphous, top free, 7 mm.-1 cm. long; style white, light purple at the top; ovary 3-4 mm. ht., 3 mm. diam., stigma capitate, 3 mm. diam.; main flowering season February to March, also all they year round.

*External fruit characters.* Shape spherical, size 4.5-5.3 cm. ht., 4.5-5.1 cm. diam., wt. 1.5-2.5 oz.; D/H index average 1.04; colour light greenish yellow; surface very smooth; oil glands in flush with the skin, also finely pitted, glossy, stem, 1-1.4 cm. long; base evenly rounded; calyx, thin; apex very minutely mamillate; areole with irregular circular depression, areolar area ribbed, slightly rugose diameter of the areolar area 1-1.5 cm., stylar scar occasionally even, usually depressed in a pit, style not persistent.

*Internal fruit characters.* Rind very thin, thickness (fresh fruit) 1-2 mm., firmness leathery, adherence strong, oil glands small, number 32-35 per ¼ sq. cm., aroma distinctly of lime, pleasant; mesocarp (albedo) white, thin, slightly sweet or tasteless; axis solid, diameter 5 mm; segments 9-10, adherence strong; pulp-vesicles slender, usually fusiform, tips acute, melting, greenish white, 0.5-1.1 cm. long; juice plenty, ½-1 oz., sour, flavour excellent; rag little; seeds small somewhat ovoid, sides plano-convex, micropylar end pointed, number few, 9-14 per fruit, size usually 1-1.1 cm. long, 7-8 mm. broad, 5 mm. thick, testa white, tegmen light brown, chalazal spot dark brown, cotyledon light greenish white.

The Karimganj lime (round variety) is very much popular in the district of Cachar where it is grown in the kitchen gardens almost in every home. The scope for its extensive cultivation in Assam is undoubtedly very great. This is a very heavy fruiter and more than a thousand fruit per tree in the peak season is not unusual. The variety is comparatively less susceptible to scab, canker and gummosis than the *Abhayapuri* lime referred to earlier. The propagation is similar to the *Abhayapuri* lime.

## The Sweets Limes (*Citrus limetta* (Risso) Lushington)

*Origin and name.* Wight (1849-50) reported sweet lime to be growing wild in the Nilgiris and he considered the variety to be indigenous to India. That it was commonly cultivated in India was stated by Roxburgh (1832) and Brandis (1874). The mention of *amrat-phal* in Babar's Memoirs, considered by many to be a form of the sweet lime, lends further prop to the above contention.

In Assam, the variety is of sporadic occurrence, growing both in the hills and in the plains. Specimens have been collected from the inaccessible interior of the Lushai Hills and of the North Cachar Hills and also from many places in the plain districts.

The variety is known as *mitha-kaghzi* or *mou-muri* in Assamese (*mou* meaning honey), *soh-bakhlein* in Khasi (*bakhlei* meaning tasteful) and *mitha (sweet)* or sharbattee in trade *(sherbet)* meaning a sweet drink), the various names indicating the sweet taste of the fruit.

That it is an introduced fruit in Assam seems to be almost certain as otherwise this variety with its tasty and likeable fruits would have reasonably a wider distribution in the province.

*Nomenclature.* The specific status of the so called sweet lime has not been recognized by Swingle (1946). Webber (1946), while admitting the morphological difference between the sour lime and the sweet lime, included it as a variety under the species *C. aurantifolia,* Swingle (sour lime). He also suspected the sweet lime to be of doubtful hybrid origin. Roxburgh (1832/1874) dealt sweet lime under the species *C. acida* while Hooker (1872) gave it a varietal recognition, although under a separate species and named it as *C. medica* var. *limetta.* Brandis (1906) included it under the same species as was done by Hooker without assigning to it a verietal name, though he (1874) sought to differentiate the sweet lime of India by referring *C. limetta* as a name for the same. The sweet lime had been named botanically as *C. limetta* as distinct from *C. lumia* (sweet lemon) by Risso and Poiteau (1818). Lushington (1910) accepted the name *C.limetta* for sweet lime of India. Tanaka (1937), a however, pointed out that *C.limetta* of Risso was quite different from *mitha* of Lahore (Punjab), *sakhar-nimbu* of Poona (Bombay) and placed the *mitha-nimbu* under a separate specific name *C.limettioides.,* Lal Sing and Sham Singh (1942) identified sweet lime (*mitha* of the Punjab) as *C.aurantifolia* var. Swingle. They possibly followed Hatton (1932) for the nomenclature of the sweet lime. It may be stated here that the *mitha-kaghzi of* Assam is morphologically indistinguishable from the *mitha* of the Punjab both of which have been studied side by side at the Citrus Experiment station, Burnihat. Javaraya (1943) while giving a digest of the *Citrus* classification in the *Marketing Report of the Citrus Fruits of India* covered the sweet lime and the sweet lemon under one botanical name *C. limonia* var. *limetta* Hort. From the above it will be apparent that most of the old writers accept the existence of sweet lime in India and seek to give some botanical name to distinguish it from the so called sour lime *(C. aurantifolia).* In order to bring out vividly the difference between these two well defined and well known fruits of India, the main distinguishing morphological characters for the both are furnished below.

| Sour lime | Sweet lime |
|---|---|
| 1. Plant usually bushy with slender branches. | 1. Plant usually medium-sized with thicker branches. |
| 2. Leaf small with prominent and distinct petiolar wing. | 2. Leaf medium to large with almost no petiolar wing. |
| 3. Leaf colour deep green or green. | 3. Leaf colour distinctly very light yellowish green. |
| 4. Crushed leaf aroma distinctly of the true sour Lime. | 4. Crushed leaf aroma distinctly different from the true sour lime. |
| 5. Flower small, usually racemose, white, very rarely slightly tinged but ephemeral, stamen 20-28, flowers all the year round. | 5. Flower large, cymose, pure white, never tinged, stamens 20-30, flowers once annually. |
| 6. Fruit very small with diam. 4-5 cm. (1½-2 in.). | 6. Fruit usually bigger than the Sour lime, diam. 6.8-7.5 cm. (2¼-3 in). |
| 7. Rind very thin, 1-3 mm. thickness. | 7. Rind thicker than the true sour lime, 3-4 mm. thickness. |
| 8. Axis solid, never semi-hollow. | 8. Axis usually semi-hollow, rarely solid. |
| 9. Pulp-vesicles intensely sour. | 9. Pulp-vesicles sweet. |
| 10. Vesicles grenish white in colour. | 10. Vesicles pure white in colour. |
| 11. Seed small. | 11. Seed always larger than that of sour lime. |
| 12. Tegmen colour light brown. | 12. Tegmen colour ivory white. |
| 13. Chalazal spot dark brown. | 13. Chalazal spot yellow ochre. |
| 14. Cotyledon greenish white. | 14. Cotyledon ivory white. |

Foregoing distinctive characters of the sweet lime and the sour lime clearly suggest that there is no justification for grouping them together as was done by many taxonomists. Besides, there is nothing to be gained by stigmatising the sweet lime to be of hybrid origin unless it is proved by actual crossing that it is really so.

In view of the antiquity of its existence and of the distinctiveness of its characters, it is very much desirable that a specific rank be assgined to it to separate it form the sour lime (*C. aurantifolia*). Of all the botanical names of the sweet lime, *C. limetta* Risso, being the oldest, should be retained.

## Key to Variery of the Sweet Lime

A. Flower once in a year, pure white.

B. Plant medium tall; spreading; moderately dense foliage; thorny; leaf colour yellowish green; petiole wingless, veins prominent on upper side.

C. Fruit almost spherical; size 7-7.8 cm. ht., 6.8-7.5 cm. diam.; surface very smooth, glossy; light lemon yellow, apex almost rounded to very slightly nippled, rind thin, axis hollow; pulp-vesicles white, insipid sweet; seeds few, chalazal cap light yellow ochre.

$V_1$.......*Mitha-kaghzi.*

### *Mitha-kaghzi*

*Plant characters.* Plant medium tall, 12-16 ft. ht, spreading; moderate-by dense foliaged; thorny, 1-4.5 cm. long, slender; prolific bearer; fruiting season middle of September to December, one crop annually.

*Leaf characters.* Leaf medium-sized; shape ovate-oblong; size 9.5-12 cm. long, 5.1-7.7 cm. broad, apex obtuse, margin crenate; petiole with rudimentary wing or wingless, petiolar wing 1-1.2 cm. long, 5 mm. broad; medium coriaceous light yellowish green colour; veins slightly prominent on upper surface; crushed leaf aroma distinct, unlike sour lime.

**Mitha-kaghzi (*C. limetta*)**

*Flower character.* Staminate and bisexual, mostly bisexual, glomerulate 2-12, cymes; pedicel 4-5 mm. long, peduncle very short, 4-6 mm. long, glabrous; calyx, deltoid, 4-5 lobed, apex acute; corolla imbricate petals 4-5 usually 5, colour pure white on both sides, size 2-2.5 cm. long, 6-9 mm. broad, shape oblong, reflexed, punctate with pellucid dots on outside; stamens 20-32, anthers 3-5 mm. long, filaments white, jointed by 2-4, some free, 0-7-1.5 cm. long; ovary 4 mm. ht., 3 mm. diam., style cylindrical, 1-1.2 cm. long including stigma,

**Soh-synteng (C. limon) (left), and Mitha-kaghzi (C. limetta) (right) a comparative study.**

greenish white colour, stigma capitate, 3 mm. diam., flowering season February to March.

*External fruit characters.* Shape usually almost spherical or slightly globose, size 7-7.8 cm. ht., 6.8-7.5 cm. diam.; wt. 5-7 oz., D/H index 0.88-1.02 (av. 0.95); colour light yellow, surface very smooth and glossy, section showing through the skin, oil glands in flush with the skin, rarely finely pitted; base evenly rounded; calyx small, thin, tip acute; apex either obtuse or very slightly nippled; areole indistinct, very minutely marked with irregular circular furrow, areolar area usually smooth, diameter 1.5-1.8 cm. stylar scar even style not persistent.

*Internal fruit characters.* Rind thin, thickness 3-4 mm. (in fresh fruit) firmness leathery, adherence medium to somewhat strong; oil glands number 30-38 per ¼ sq. cm., shape globose to oblate, size very small, glandular layer about 1/3 rind, aroma distinct; mesocarp (albedo) white thin, soft, sweet; axis usually semi-hollow or rarely solid, diam. at the centre 0.8-1.5 cm.; segments 10-11. adherence medium, septa thin, white, slightly tender, dorsal contour truncate to slight convex, carpel size 6-6.8 cm. long, 2.5-3 cm. broad, 1-2 cm. thick; pulp-vesicles uniformly white, fusiform, 8 mm.-2.2 cm. long, 2-4 mm. wide, stalked, texture medium coarse, somewhat soft and melting, tips usually attenuated and obtuse, arising from dorsal side; juice plenty, 1.5-4 oz., taste sweet or insipid sweet, flavour agreeable; rag medium; seeds few, 6-10 per fruit, size 1.2-1.6 cm. long, 7-8 mm. broad, 5 mm. thick, shape ovate, slightly beaked or pointed at micropylar end, very minute anastomosing lines on one side, other side usually with a small smooth area, testa cream white, chalazal spot light yellow ochre, cotyledon white or ivory white.

The *mitha-kaghzi* is not very common in Assam and is generally grown as a novelty in homestead gardens to cater to the needs of the family. The similarity of the *mitha-kaghzi* of Assam with the variety Palestine as described by Webber (1946) is very apparent. Of late people of this province have developed a fancy for this fruit perhaps due to its early season (September to November) when no other local sweet citrus fruits or oranges are available.

The propagation of this variety is usually done by *gootee* or marcettage and occasionally nucellar seedlings are also planted. The *soh-myndong* (rough lemon) has proved to be compatible as a stock for the *mitha-kaghzi* in Assam. It has not been tried as a stock for the commercial citrus varieties of the province.

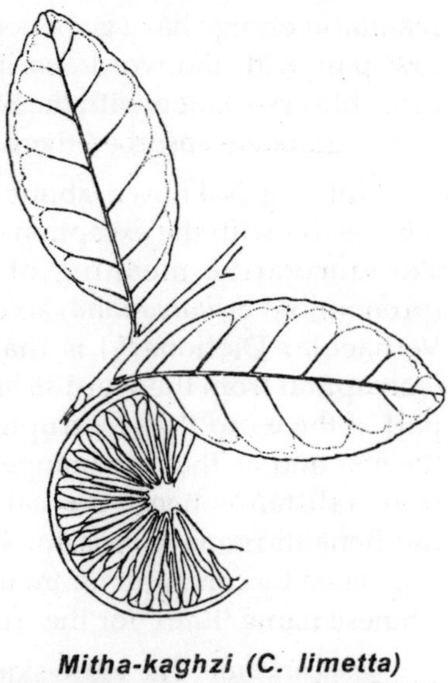

***Mitha-kaghzi* (*C. limetta*) showing the outline of leaf and internal characters of fruit.**

## The Loose-skinned Mandarin Oranges (*Citrus reticulata* Blanco)

*Origin and name.* Most of the taxonomists seem to agree that Cochin China, the southern most region of the Indo-Chinese Peninsula is the birth place of the mandarin orange (Tolkowsky, 1938). Brandis (1906) reports that the mandarin orange was found wild in Garhwal (Thomson,) Kumaon (Strachey), Sikkim and the Khasi Hills (Hooker), Manipur (Watt) and mountain forests in Peninsula (Beddome). The Khasi orange (*C. reticulata* Blanco), as the mandarin is known in Assam, has been in existence in this province from a hoary past, mostly in the montane and submontane tracts. Whether it is indigenous to Assam or it was introduced, seems to be a matter of conjecture as no historical evidences are available to support one view or the other. If it was introduced it might have travelled to Assam in a period of silent barter, known to have been in existence between the hill people bordering Yun-nan province of China, Upper Burma and Assam. As from the eighth century, the Shans of the Yunnan province had carried out a series of invasions in the Brahmaputra valley and completely occupied the area in the thirteenth Century (McMahon, 1886). It is quite likely that the Shans who later settled on the land and took to agriculture might have brought with them fruits or seeds of this luscious fruit and planted the same with success in their adopted country and the variety thus became gradually feral in Assam.

The Khasi name *soh-niamtra*, the Assamese name *sumthira* and the common Indian name *sangtra, santara. santra* or *suntara* for the mandarin orange have got a perceptible phonetic semblance in the first part with the word *San* indicating a common origin and a probable association with the Shans that invaded Assam and who were ultimately absorbed there.

Nothing is known about the meaning of the Khasi name *soh-niamtra* with the exception of the word *soh* which means fruit. The derivative meaning of the Assamese word *sum-thira* (pronounced as *humo-thira*) according to the Hem-kosh (Assamese Vernacular Dictionary) is that the first part of the word sum (corruption from the word *shon*-gold) means gold and the second part of the word *thira* (corruption from the word *tara-star*) means the star and as the ripe orange in the tree looks like golden stars from a distance, hence it is called *sum-thira.* It is quite likely that the Bengali name *kamala* for loose skinned orange might have originated by corruption from the Burmese name *kambala-thi* or the Chinese name 'Kam' for the fruit indicating a foreign origin.

*Nomenclature.* The loose-skinned mandarin oranges of India do not seem to have received so far the attention that they deserve from the systematists. The mandarin orange is *par excellence* the most important commercial fruit variety of India and yet it has not been given any separate specific status by Roxburgh (1832), Hooker (1872) and Brandis (1874 and 1906), all of whom merely recognized it as a variety under the species *Citrus aurantium* Linn. Bonavia (1890) sought to separate out the mandarin orange from the so called sweet orange or Malta type and suggested the name *Citrus curantium sinensis* Rumphius for the former and *Citrus aurantium sinensis* gallesio for the latter but the distinction between the tight-skinned and the loose-skinned orange is so great that in no way these two can be grouped together under the same species *Citrus aurantium* Linn. which now stands for the seville or bitter orange, a fruit distinctly different again from the sweet orange (*C. sinensis).*

Lushington (1910) was probably the first systematist in India to raise the mandarin orange to a specific rank and called it *Citrus chrysocarpa* which later had the approval of Tanaka (1937, a). But Tanaka's delineation of the mandarin oranges (1937, b) went probably too far, as he placed it under a subgenus *Meta-citrus* (solitary flowers) splitting it further into three sections *Osmo citrus* (*C. junos Sieb*), *Micro carpa* (*C. mitis*) and *Acrumen* the true mandarin, the last one was sub-divided again into two sub-sections namely *Euacrumen* (*C. nobilis*) and *Micro acrumen* (*C. chrysocarpa*). It will thus appear that Tanaka has assigned a specific status to almost one and all mandarin types of oranges belonging to each section or sub-section of the sub-genus *Meta-citrus;* although the fruits of each section may conveniently be treated as a variety within the subgenus without basically altering the structure of the mandarin group. He, however, emphasises the need of a standardized scientific name at least for Indian *santara* (mandarin).

Writing in Bailey's *Standard Cyclopedia of Horticulture,* Swingle (1919) recognized *C. nobilis* Lour (the so called King orange) as the specific name for mandarin oranges and considered true mandarin as var. *deliciosa*. But in a recent publication Swingle (1946) points out that the true mandarin (loosely attached rind and carpel) is very different from *C. nobilis* which Loureiro (1790) described in his *Flora of Cochin-China*. Swingle also states, quoting Matlack (1931) that the chromatoplasts of *Citrus nobilis* (King orange are different from those of the loose-skinned mandarin orange. Thus it is clear that *C. nobilis,* hitherto, accepted as a specific name for mandarin type of oranges cannot justifiably stand for the group. The oldest name of the mandarin oranges as pointed out by Swingle (1946) is *Citrus reticulata* which was described by Blanco in 1837, in his *Flora of the Philippines*.

Subsequently, Merril's (1918) study on Blanco's plants reveals clearly that *C. reticulata* as described by Blanco correctly fits in with the true mandarin. Swingle (1946) having been convinced of this, accepts *Citrus reticulata* as the vaid name for all mandarin oranges. In the present state of confusion of nomenclature, it is desirable to follow Swingle till a critical study of the mandarin group as a whole justifies a different specific name other than *C. reticulata*.

Webber (1946) classified real mandarins (*C. reticulata*) into five groups, namely, King Satsuma, mandarin, Tangerine and Mitis; although later he considered 'King orange' to be a hybrid and not as a type of the species. A broad distinction has also been made by him between the red and the yellow coloured fruits and he has termed them as Tangerine and Mandarin respectively. These two terms are well established in trade in America and other Western countries.

The group Satsuma which constitutes a distinct class especially in Japan has not been met with in Assam while Mitis group is represented by only one variety here. Hume (1941) placed all the fruits belonging to the above groups except Mitis within the mandarin group and considered them separately as varieties.

The present writers, finding all these different groups very closely related taxonomically to the true mandarins, assign to them individually a varietal status and accordingly all the fruits of the group (mandarin) have been keyed for as varieties.

The mandarin oranges of Assam may be seggregated into four distinct groups on the basis of (i) smoothness or roughness of skin, (ii) red or yellow-orange colour of the rind, (iii) size of fruits, and (iv) acidity and sweetness of pulp. But these diagnostic characters are reckoned with as varietal manifestations and not specific.

## Key to Varieties of the Khasi Mandarins

A. Fruits with sweet pulp (sweetness and acidity well blended).

B. Tree medium to tall. erect; densely foliaged; leaf deep green; thorny to almost thornless; flower solitary, small to medium, pure white, annual bloom.

C. Fruit medium-sized, (7-9 cm. diam.) shape depressed globose, surface, orange yellow to deep orange, smooth, rind thin, very loosely attached; carpels easily separable;

aroma distinct; seeds somewhat clavate, chalazal spot brown, cotyledon green, polyembryonic.

$V_1$................*Soh-niamtra.*

CC. Fruit medium-sized, (6.5-8.5 cm. diam.) shape oblate; surface yellow to light orange, somewhat rough; rind usually thicker than that of *soh-niamtra.* loosely attached; aroma peculiarly aromatic; carpel easily separable; seeds obovate, chalazal spot brown, cotyledon greenish white.

$V_2$..............*Soh-umkdai.*

CCC. Fruit very small (3.5-4 cm. diam.), almost globose to oblate; surface smooth, light orange colour; rind very thin, loosely attached; aroma like that of *soh-niamtra;* carpel easily separable; seeds small, turbinate, chalazal cap sepia-brown, cotyledon light green.

$V_3$.................*Naga-santra.*

AA. Fruits with acid pulp.

B. Tree small to medium, erect; densely foliaged, leaf deep green; usually thornless; flowers solitary, small, pure white, annual bloom.

C. Fruit small to medium-sized (4-6 cm. diam.) oblate; surface smooth. scarlet : rind thin, loosely attached; carpel easily separable; seeas small turbinate to clavate. chalazal spot deep brown, cotyledon light

$V_4$..............*Soh-siem.*

CC. Fruit small to medium (6-7 cm. diam.), oblate; very warty; colour

$V_5$..........*Kapura-tenga.*

BB. Tree small to medium, erect; densely foliaged, leaf small, crushed leaf aroma approaching resinous smell, light green colour : flower very small, pure white, flowers all the year round.

C. Fruit very small (3-4 cm. diam.), globose; surface smooth, deep orange colour : rind thin, loosely attached; aroma distinct, approaching kamquat; carpel separable; seeds small, almost orbicular, chalazal cap brown, cotyledon light green.

*Soh-niamtra*

**(Top) *Soh-niamtra* (*C. reticulata*) grown at Shella (Khasi Hills); (Bottom) *Soh-niamtra* (*C. reticulata*) grown at Burnihat. Shape and size compared**

*Plant characters.* Plant medium to tall, ht., 15-25 ft. erect habit, branches erect in young trees, spreading in old trees, well branched : very densely foliaged; both thorny and thornless, thorns usually confined to trunk and lower branches, growing shoots angular, glabrous; plants (seedlings) somewhat slow growing, hardy; fruiting season usually November to February, one crop in a year, prolific bearer, trees with recorded yield of about 4500 fruits per tree sometimes met with.

*Leaf characters.* Leaf medium-sized, petiole rudimentary or wingless shape broad-lanceolate, size small ones usually 5.5-7.5 cm. long, 2-3.2 cm. broad with petiole 8 mm.-1 cm. long, normal ones 8.5-11.5 cm. long, 4-6.2 cm. broad with petiole 1-1.5 cm. long, 1.2 mm. broad, big ones usually 12-12.4 cm. long, 6-6.5 cm. broad with petiole 2.2-2.5 cm. long, 4 mm. broad; margin crenate, apex attenuated and almost acute, emarginate, medium coriaceous, dark green, lighter beneath; veins slightly prominent on upper surface, articulated.

*Flower characters.* Bisexual only, solitary, rarely in pairs; pedicel 2-3 mm. long, peduncle indistinct, glabrous; calyx thin, deltoid, 4-5 lobed, apex acute, ciliate, colour greenish white; corolla imbricate, petals 5, colour pure white, oblong, apex almost acute, size usually, 1-1.4 cm. long, 5-8 mm. broad; stamens 14-19, (usually 16-17), filament white, glabrous, polyadelphous, jointed by 2-3 together, occasionally 1 or 2 free, 6-9 mm. long, anther 1-2 mm. long; ovary globular to oblate, 1.5-2 mm. long, 2 mm. broad, 8-13 loculi, style terete, greenish white, 6-7 mm. long including stigma; flowering season February to April.

*External fruit characters.* Shape depressed globose to oblate, occasionally turbinate, section showing through the rind, size big ones 7.4-7.8 cm. ht., 8.8-9.1 cm. diam., normal ones 6-7 cm. ht., 6-8 cm. diam.; wt. 5-10 oz. (usually 5-8 oz.); D/H index 1.07 to 1.26 per cent (av. 1.20); colour orange yellow to bright orange; surface smooth, glossy; oil glands usually in flush with the skin, also minutely pitted, rarely slightly papillate; base usually even and obtuse, occasionally with very low neck, slightly ribbed; calyx small, thin, 4-5 lobed, tip usually acute; stem short, 5-6 mm. long; apex broadly depressed or retuse, usually forming a broad apical sinus; areolar area very slightly rough, pitted, navel very rarely present, if present protruded into very small nipple, stylar scar present, diameter 2-3 mm., even, style not persistent.

*Internal fruit characters.* Rind thin, thickness 3-5 mm., firmness soft to very soft, adherence very slight or none; oil glands medium to small, many, number 30-55 per ¼ sq. cm., shape usually pyriform, oval or ellipsoid, size 0.5-3 mm. ht., 0.5-1.5 mm. diam., aroma distinctly of orange (mandarin), oil abundant, glandular layer conspicuous, more than ¾ or rind; mesocarp (albedo) white to rather very slightly orange tinged, soft, spongy, almost tasteless or very slightly sweet, vascular bundle prominent and reticulated over the carpels; axis hollow, diameter at base 0.8-1.1 cm., at the centre 1.5-2.4 cm., at apical end 1-1.2 cm., carpel usually not broken at axis, occasionally broken in much ripe fruits; segments 8-13 (usually 10), adherence very slight, rarples very easily separable; rag little to medium; septa very thin, soft, dorsal contour convex to slightly truncate, carpel size usually 5-6 cm. long, 2.2-2.7 cm. broad, 1.6-2.6 cm. thick; pulp vesicles uniformly orange coloured texture somewhat coarse, very melting, arising from both dorsal and radial walls, shape irregular, short and broad, polygonal, size big ones 1.2-1.6 cm. long, 3-6 mm. broad, stalked, 1-1.4 cm. long, apex usually obtuse to slightly acute, flavour agreeable; juice abundant, 3-6 oz. per fruit, colour orange, sweet or sweetness and acidity well blended; seed medium-sized, number 8-25 per fruit (usually 10-15), top shaped to somewhat clavate with distinct fin-like projection of testa towards micropylar end, surface smooth, size 1.1-1.5 cm. long, 6-8 mm. broad, 5-6 mm. thick, testa white, tegmen light brown, chalazal spot brown, cotyledon green, polyembryonic.

The *soh-niamtra* is sometimes called *soh-thiang* (*soh* means fruit; *thiang* means sweet) and also *soh-bang* (bang meaning sweet) by the Khasis. This is a mandarin type of arange *par excellence*, grown extensively all over Assam as a commercial crop and is destined to retain its present status throughout the whole future of the citrus industry of the province. Its wide range of adaptability for diverse soils and climatic conditions is well known. The estimated area under this variety is about 20,000 acres (1948). The main centres of production are in the districts of the Khasi Hills (southern and northern slopes), Kamrup, Goalpara, Darrang, Sibsagar, Cachar and Lushai Hills. There are also minor areas, virtually spread all over the hilly terrains of the province. The fruits of the southern slopes of the Khasi Hills are considered to be the best in whole of Assam. There is a late ripening variety growing in Mawryngkneng side (4000-5000 ft. altitude), a locality towards

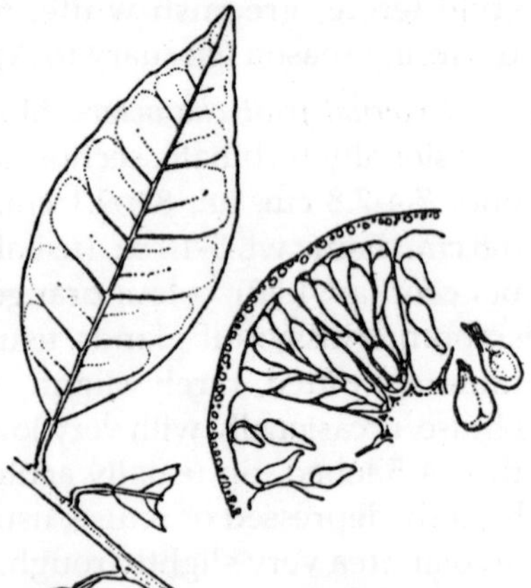

***Soh-niamtra (C. reticulata)* showing the outline of leaf and internal structure of fruit**

Jowai, about 20 miles east of Shillong, the capital of Assam. The fruits of this variety generally come in Shillong market from April to early part of May when no other oranges are available. It has been found that this character of lateness is due to the environmental causes as both nucellar seedlings and budded progenies of the same do not exhibt the similar behaviour under altered soil and climatic conditions of Burnihat. Sweet mandarin oranges of Assam show remarkable differences in size, shape, colour, taste and thickness of rind of fruits under different soil and climatic situations, so much so that in trade many of these pass for as varieties named after the respective places of their production, e.g. Shella orange, Burnihat orange, Jatinga orange, Lushai orange, Karimaganj orange etc. But from a prolonged study of both nucellar and clonal progenies of the same it has been ascertained beyond doubt that these differences are primarily due to soil and climatic factors and are not tenable from taxenomic points of view as the different characters referred to above have been found to change with the change in environment. The authors, therefore, strongly hold the view that the so-called sweet mandrain oranges of Assam are all one and the same irrespective of the place of their occurrence and as such they refrain from giving them individually a varietal recognition.

***Soh-umkdait***

**Soh-umkdait (*C. reticulata*)**

*Plant characters.* A medium tall tree, ht. 15-20 ft., up right, densely foliaged, usually thornless, thorn 1-3 cm. log; fruiting season December to March, one crop annually.

*Leaf characters.* Medium-sized, petiole almost wingless, or rudimentary wing, shape broad lanceolate, size 5.5-11 cm. long, 3-5.2 cm. broad, petiole 0.8-2 cm. long margin crenate, apex attenuated and somewhat acute, emarginate, medium coriaceous, light green, articulated, crushed leaf aroma somewhat aromatic.

*Flower characters.* Bisexual only, usually solitary, occasionally in pairs, pedicel 3-4 mm. long, peduncle indistinct, glabrous; calyx thin, stellate, deltiod, apex acute, ciliate; corolla imbricate, petals 4-5, oblong, pure white on both sides, 1.2-1.7 cm. long, 4-8 mm. broad; stamens 16-18, filament white, glabrous, usually free, rerely jointed by two, 0.8-1.1 cm. long; anther 1-2 mm. long; ovary small, 1.5-2 mm. ht., 2 mm. diam., oblate, style terete, greenish white, glabrous, 7-9 mm. long including stigma, stigma capitate, diam. 2 mm.; flowering season February to April.

*External fruit characters.* Shape oblate, size 5-6.1 cm. ht., 6.5-8.5 cm. diam.; wt. 3-7 oz.; D/H index 1.22 to 1.50 (av. 1.35); colour orange yellow to orange, surface slightly warty, rather somewhat pebbled, glossy, oil glands either in flush or slightly pitted; base usually depressed or even, rarely with low neck, ribbed, basal area rough; calyx depressed in base, division indistinct, tips blunt; stem short, 6-8 mm. long, strongly attached; apex broadly depressed, forming a broad apical sinus; areole absent, areolar area more or less warty, stylar scar present, usually even, or depressed in a small apical pit, style not persistent.

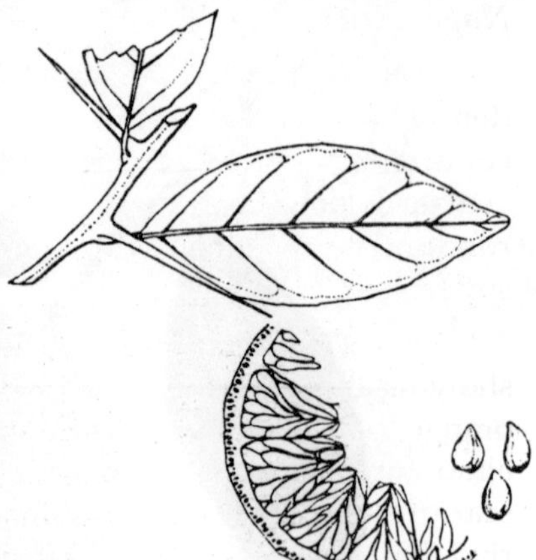

**Soh-umkdait (*C. reticulata*) showing outline of leaf and internal structure of fruit.**

*Internal fruit characters.* Rind somewhat thick, thickness 4-5 mm., firmness medium soft to soft, adherence very slight, or none; oil glands medium to small, many, number 50-58 per ¼ sq. cm., shape usually pyriform to ellipsoid, size 1-3 mm. long, 0.5-2 mm. broad, aroma somewhat spicy, distinct oil abundant, glandular layer thick, about ¾ rind; mesocarp (albedo) very thin, white, spongy, almost tasteless or very slightly sweet, vascular bundle prominent reticulated over the carpels; axis hollow, stellate in outline at median cross section, basal diam. 5-6 mm., mid-diam. 1.5-2 cm., apical diam. 0.6-1.2 cm., carpel usually closed at axis, occasionally broken; segments 8-11 (usually 9-10), adherence very slight, carpels easily separable; rag little; septa thin, slightly tough, dorsal contour convex to slightly truncate, carpel size 3.5-5 cm. long, 1.8-2.5 cm. broad, 1.5-2.5 cm. thick; pulp-vesicels light orange coloured, texture somewhat course and loose, not very melting pulp-vesicles arising from dorsal and radial walls, shape irregular, ovate-oblong, short and broad, polygonal, tips both acute and obtuse, 0.5-1.5 cm. long, 3-4 mm. broad, flavour good, distinctly spicy; juice plenty, 2/4 oz. per fruit, sweetness and acidity blended; seeds 8-20 (usually 10-12), size 1-1.2 cm. long, 6-7 mm. broad, 5 mm. thick, shape somewhat obovate, or D-shaped, micropylar end pointed, one side almost smooth, other side with minute ridges, rough at rounded end, testa white, tegmen light brown, chalazal spot brown, cotyledon light green.

This variety is distinctive by its characteristic spicy flavour of its rind. It is generally met with in the Khasi Hills. At present it has no commercial value beyond local importance. The fruit is quite good and keeps well due perhaps to its slightly thicker rind. It is a novelty fruit grown usually in home-stead gardens.

### *Naga-santra*

*Plant characters.* A medium tall plant, ht. 10-12 ft. upright habit, densely foliaged, almost thornless; leaf broad lanceolate, medium coriaceous, light green, apex attenuated, obtuse, margin slightly crenate to almost entire, size 4.5-6.5 cm. long, 2.5-3.5 cm. broad with rundimentary petiolar wing (flower not seen); one crop in a year, shy fruiter.

*Fruit characters.* Shape spherical to slightly oblate, section showing through rind, size 2.8-4 cm. ht., 3.5-4 cm. diam., colour orange to deep orange, surface smooth, very minutely pitted; base almost obtuse; areole absent, areolar area very slightly rough, stylar scar present, depressed in a small apical pit, style not persistent; rind thin, thickness 3-4 mm.; oil glands conspicuous, size 0.5-1 mm. long, 0.5-1 mm. broad, shape spherical or slightly oblate; mesocarp (albedo) orange coloured, sweet; axis hollow, mind diam. 0.8-1 cm.;

segments 8-10, adherence slight, carpels easily separable, dorsal contour convex to slightly truncate; rage little; pulp-vesicle colour orange, texture coarse, small, polygonal, tips blunt, melting; juice scanty, sweetness and acidity well blended; seeds very few, 3-8 per fruit, almost top-shaped, size 0.9-1 cm. long, 5 mm. broad, 4 mm. thick, surface almost smooth, slightly dented towards rounded end, testa cream white, tegmen light brown, chalazal cap sepia-brown, cotyledon light green.

This is a very small sized fruit found really in a wild state in far off forests of Mokekchung (Naga Hills) at an altitude of about 4000 ft. So far ascertained, this variety of mandarin orange has never been met with either in the villages or in the regular plantations. The fruits being very small, though somewhat sweet, have hardly any commercial importance. It may not be surprising that this interesting variety on further examination might prove to be the progenitor or the primitive ancestor of the so-called *soh-niamtra* (Khasi mandarin) of Assam. This requires further investigation.

**Soh-siem (*C.reticulata*)**

### *Soh-siem*

*Plant characters.* A medium sized plant, ht. 15-18 ft., upright and erect habit, densely foliaged, usually thornless, thorn 1-5 cm. long, young shoot angular, glabrous; fruiting season November to January, one crop annually.

*Leaf characters.* Medium-sized, shape oblong-elliptical to broad-lancealate, size 5.5-9 cm. long, 2.5-4.5 cm. broad, petiole 0.8-1.3 cm. long, margin crenate (usually top half crenate and lower half almost entire), apex attenuated and obtuse, emarginate, medium coriaceous, dark green, articulated.

*Flower characters.* Bi-sexual only, usually solitary, occasionally in pairs; pedicel short, 2-3 mm. long, peduncle 5-6 mm. long, glabrous; calyx deltoid; apex acute, ciliate, 4-5 lobed; corolla imbricate, petals 4-5, oblong, pure white, size 0.8-1.2 cm. long,

3-5 mm. broad; stamens 15-19, filament white glabrous, polyadelphous, jointed by 3-7 together, also 1 or 2 free, 5-8 mm. long, anther small, 1 mm. long; ovary usually 1 mm. long, 1.5 mm. broad, oblate, 7-12 loculi, style white, terete, glabrous, 9 mm. long including stigma, stigma capitate; flowering season late February-April.

*External fruit characters.* Shape depressed globose or oblate, size normal ones 3.3-4.3 cm. ht., 4-5.8 cm. diam., big ones (rare) 5.2-5.4 cm. ht., 6-6.5 cm. diam., wt. 1¼-4½ oz. (usually 1¼-2¼ oz.); D/H index 1.21-1.38 (av. 1.30); colour scarlet or vermilion; surface smooth, minutely foveolated glossy; base depressed, or somewhat even, ribbed, furrows extending 1-2 cm.; calyx small, stellate, apex acute, 4-5 lobed; stem small, 5 mm. long, strongly attached; apex depressed with an apical sinus; areole absent, areolar area almost smooth, finely pitted, occasionally ribbed, navel very rarely present, very small stylar scar present, depressed in a small apical pit, style not persistent.

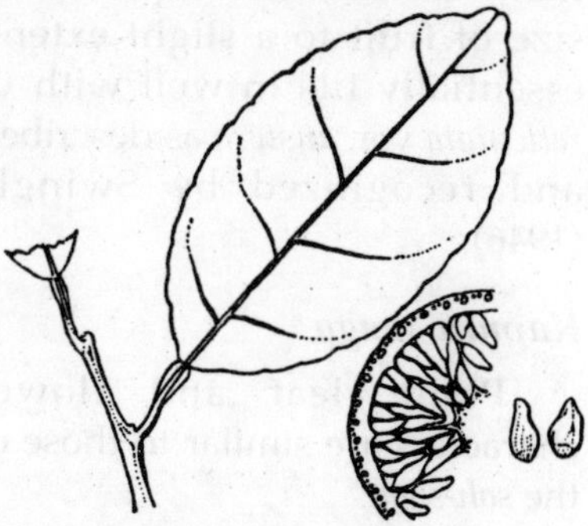

**Soh-seim (C. reticulata) showing outline of leaf and internal structure of fruit.**

*Internal fruit characters.* Rind thin, thickness 3-4 mm., very soft, adherence very slight or none, slightly puffing; oil glands number 50-58 per ¼ sq. cm., shape spherical, oval or slightly pyriform, size 0.5-1.5 mm. long, 0.5-1 mm. broad, oil abundant, aroma similar to the Khasi mandarin, glandular layer abcut ¾ rind; mesocarp (albedo) almost white with red tinge, spongy, almost tasteless or very slightly sweet; vascular bundle prominent, reticulated over the carpels; axis hollow, basal diam. 5-8 mm., mid diam. 0.8-1.1 cm., apical diam. 6-8 mm., stellate in outline in median corss section, carpel rarely broken at axis; segments 7-12 (usually 9-10), adherence slight; rag little; septa very thin, slightly tough, dorsal contour convex to ver slightly truncate, carpel size 2.8-3.1 cm. long, 1.5-1.6 cm. broad, 1.2-1.4 cm. thick; pulp-vesicles deep orange colour, texture coarse, vesicles arising from dorsal and radial walls, very melting, shape ovate or slightly fusiform, polygonal, tips both acute and obtuse, size 0.5-1.1 cm. long, 3.4 mm. broad; juice medium, ¾ to 1 oz. per fruit, colour deep orange, very sour; seeds several, 10-24 (usually 12-18) per fruit, small, size 1-1.2 cm. long, 5-6 mm. broad, 4-5 mm.

thick, top-shaped, one end obtuse, other end pointed with fin-like projection of testa, surface almost smooth, testa white, tegmen light brown, chalazal spot dark brown, otyledom light green.

The *Soh-siem* is very common in the Khasi Hills. The fruit is very attractive due to its vermillion or scarlet colour of the rind and is usually used for decoration of houses during festivals. The name *soh-siem* in Khasi language means the 'King of fruit' (*soh*=fruit and *siem* =King). this attribute is perhaps due to its attractive red colour, as otherwise the fruit is practically useless for its sourness.

Tanaka (1937, a) refers to this fruit as almost identical with *Citrus erythrosa* Hort., ex Tanaka (1927) and names it as var. *assamica* under the same species. On careful examination of the botanical characters it appears that this variety, except for the size of fruit to a slight extent, essentially fits in well with C. *reticulata* var. *austera* as described and recognized by Swingle (1946).

**Kapura-tenga (*C. reticalata*)**

### *Kapura-tenga*

Plant, leaf and flower characters are similar to those of the *soh-siem*.

*External fruit characters.* Shape oblate or depressed globose, never pyriform, size 3.5-4.5 cm. ht., 6-7.3 cm. diam., wt. 2-4.5 oz.; D/H index 1.33 to 1.77 (av. 1.57); colour vermilion or scarlet; surface very warty, coarsely pitted; base slightly depressed, prominently ribbed or furrows extending 1-3 cm.; calyx small, sunken, irregularly divided, tips obtuse; stem short, 5-8 mm. long, very strongly attached; apex depressed with broad apical sinus; areole absent, areolar area very warty, ribbed, stylar scar even in a small apical pit.

*Internal fruit characters.* Rind thin, thickness 3-4 mm., very soft, adherence very slight or none, puffing; oil glands number 45-52 per ¼ sq. cm., shape variable, usually pyriform or oval, size

2-3 mm. long, 0.5-1 mm. broad, oil abundnat, glandular layer thick, about ¾ rind; mesocarp (albedo) white with slight orange tinge, spongy, almost tasteless or slightly sweet, vascular bundle prominent, reticulated over the carpels; axis hollow, basal diam. 5-6 mm., mid diam. 1.2-1.5 cm., apical diam. 0.8-1.1 cm., carpel usually not broken at axis; segments 8-10, adherence of carpels slight, septa very thin, soft, dorsal contour convex to slightly, truncate, carpel size 3.2-3.8 cm. long, 2-2.3 cm. broad, 1.8-2.5 cm. thick; pulp-vesicles deep orange colour, texture coarse, broad and short, ovate to slightly fusiform tips obtuse arising from dorsal and radial walls, very melting, size 0.5-1.2 cm. long, 3-4 mm. broad; juice plenty, 1.5-2.5 oz. per fruit, colour deep orange, very sour; seeds, many, 20-24 per fruit, shape clavate to top-shaped, size 1.1-1.4 cm. long, 6-8 mm. broad, 4.5 mm. thick, one side obtuse micropylar end with fin-like projection of testa, surface more or less smooth with anastomosing lines on edges, testa white, tegmen brown, chalazal spot dark brown, cotyledon light green.

This variety is absolutely similar to the *soh-siem* of the Khasi Hills in all morphological details expect for its roughness of the skin. It has no commercial importance at all. The specimen has been collected from the lower Assam valley.

## *Narangi*

*Plant characters.* A small to medium sized plant, ht. 8-12 ft., upright and erect, densely foliaged, almost thronless, fruiting season all the year round.

*Leaf characters.* Leaf small to medium, coriaceous, pale green, obovate, petiole almost wingless or rundimentary, size 4.5-8.5 cm. long, 2.5-4.5 cm. broad, petiole 0.8-2 cm. long, 2-3 mm. broad, margin top half usually crenate, lower half almost entire, apex obtuse, emarginate, mid-rib prominent on both sided, crushed leaf aroma distinct, approaching that of Kumquat.

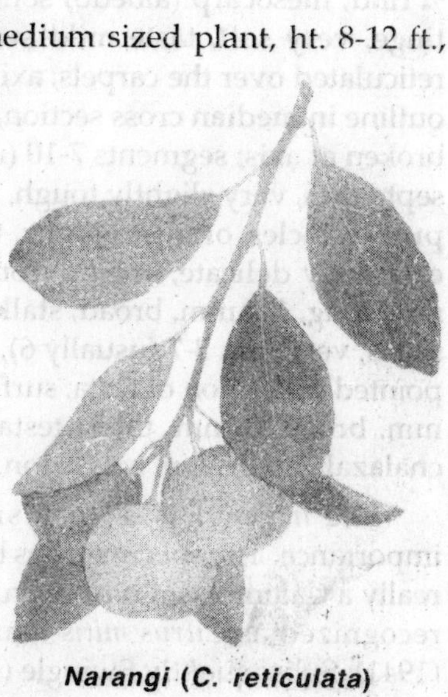

**Narangi (*C. reticulata*)**

*Flower characters.* Bisexual, usually solitary, rarely in pairs; pedicel indistinct, 2-3 mm. long, peduncle indistinct; calyx deltoid, apex acute, 4-5 lobed, margin ciliate; corolla imbricate, petals 4-5, oblong, 1.1-1,3 cm. long, 4-6 mm. broad, pure white, reflexed; stamens 20-24, filaments white, jointed 2-3 together from base upto 5 mm., polyadelphous, 1 or 2 free, glabrous, 5-8 mm. long, anther 0.5-1.3 mm. long; ovary oblate, 1 mm. ht., 1.5-2 mm. diam. 7-10 locules, style white, terete, 4-5 mm. long, including stigma, flowering season all the year round.

*External fruit characters.* Shape oblate or depressed globose, section showing through the skin, puffing, size 2.4-3.2 cm. ht., 3-4 cm. diam., wt. ½-1 oz., D/H index 1.17-1.26 (av. 1.23); colour deep orange, surface smooth, glossy, very finely pitted; base evenly rounded, smooth, very rarely minutely ribbed; calyx small, conspicuous, fitted like a small button; stem very short, 3-6 mm. long, strongly attached; apex depressed, areole usually indistinct, diam. of areolar area 7-9 mm., very slightly ribbed, stylar scar even in a small apical pit, style not persistent.

*Internal fruit characters.* Rind very thin, thickness 1-2 mm., firmness very soft, adherence very little or none; oil glands small, many, 50-60 per ¾ sq. cm., shape spherical, size slightly less than 1 mm. ht., 1 mm. diam., aroma spicy or aromatic, glandular layer ¾ rind; mesocarp (albedo) somewhat white with slightly orange tinge, very soft, taste mild sweet, vascular bundle conspicous, reticulated over the carpels; axis hollow or semi-hollow, stellate in outline in median cross section, mid diameter 6-8 mm., carpel not broken at axis; segments 7-10 (usually 8), adherence very slightly, septa thin, very slightly tough, dorsal contour convex to truncate; pulp-vesicles orange colour, texture coarse, small and broad, extremely delicate, arising from dorsal and radial walls, size 4-6 mm. long, 1-2 mm. broad, stalked, 4-6 mm. long; juicy, very sour; seeds, very few, 3-7 (usually 6), shape almost orbicular with slight pointed projection of testa, surface smooth, size 0.9-1 cm. long, 6-7 mm. broad, 5 mm. thick, testa white, tegmen very light brown, chalazal cap brown, cotyledon light green.

The *narangi* is a very small fruit with no commercial importance. The specimen has been collected from Karimganj. It is really a Calamondin type of fruit of Assam. Calamondin has been recognized as *Citrus mitis* Blanco by Swingle (1919) and Hume (1941). Subsequently Swingle (1946) considers it to be of doubtful origin. Tanaka (1937, a) included Calamondin within the mandarin group, although assigning to it a separate rank under subsection *Microcarpa*. The *narangi* of Assam bears a striking similarity with the loose-skinned orange in general characters except for the size of fruit, taste, and odour of the crushed leaf (decidedly Kumquat-like aroma) and periodical rhythm of flowering. It has therefore, been provisionally placed here as a variety under *C. reticulata* with which it seems to be closely related.

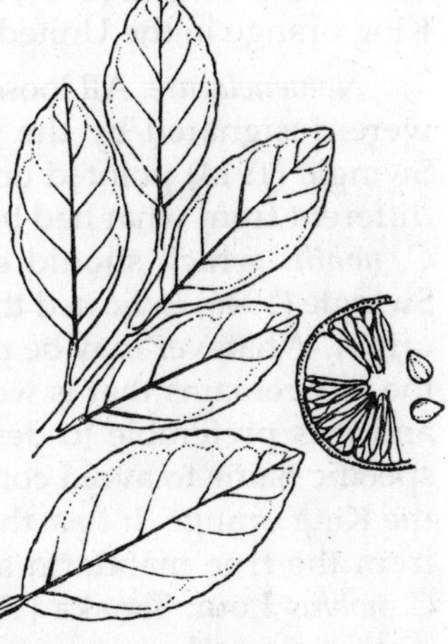

***Narangi (C. reticulata)*** **showing outline of leaf and internal character of fruit**

Bonavia (1890) pictured *narangi* like fruits under the name *kokni* orange *reshmi narangi, keonla,* etc. in Plate Nos. CXXII and CXXVII of his book. All the above named varieties, as referred to by Bonavia, have been included under one specific name *Citrus crenatifolia* by Lushington (1910).

## The King Orange (*Citrus nobilis* Loureiro)

*Origin and name.* The so-called King orange which is also known as the King of Siam,the Kunembo of Japan is exclusively designated by the name King orange in general citrus literature. It is considered by many to be a native of Cochin-China whence it has been probably introduced to other places. In Assam it is represented by the *jeneru-tenga* which is indistinguishable from the one described and illustrated under the name 'King orange' by Hume (1941) and Webber (1946). It is also fairly well represented by a drawing in Plate No. CCXXXIV of Bonavia (1890) with the only exception of the leaf petiole which is perhaps not accurately drawn. This inaccuracy of the leaf character may be due to selection of a leaf model from the stem end of the fruit where petiolar wing is usually atrophied.

Whether *jeneru-tenga* is an introduced fruit or it is indigenous to Assam, cannot be said with certainty but it is, fairly common, particularly in the plains of Assam. It is differently spelt, such as *jendru, junaru* and *jeneru,* all these different names meaning the same fruit, Tanaka (1937, a) reported to have found a form of the King orange in the United Provinces of India.

*Nomenclature.* All loose-skinned mandarin oranges previously were designated by the specific name *Citrus nobilis* Lour. But Swingle (1914) pointed out that true mandarin orange was quite different from what had been described by Loureiro (1790) under *C. nobilis* which should exclusively apply to the King orange. Swingle (1946) indicated that the so-called King orange is of hybrid origin. Whatever may be the taxonomic status of the King orange, the fact remains that is well represented in the oriental countries and it is preferable to designate this type of fruit by a separate specific name to avoid confusion between the true mandarin and the King orange. It has, therefore, been segregated by the writers from the true mandarin and has been placed under the species *C. nobilis* Lour. Tanaka (1937, a) also recognized it (*C. nobilis*) as distinct from the true mandarin (*C. reticulata* Blanco).

## Key to the variety of the king orange

A. Plant medium to tall; usually thornless, erect habit; densely foliaged, leaf deep green, coriaceous, broad-lanceolate, vein prominent, petiole spathulate; flowers bisexual only, usually solitary, stamens free, style very short.

B. Fruit medium-sized (6-8 cm. diam.), oblate, surface characteristically rough and pebbled; axis hollow; segment 10-13; pulp-short, coarse-grained arising from dorsal and radial walls sweet; seeds somewhat cuneate, chalazal spot brown, cotyledon light green.

$V_1$....*Jeneru-tenga.*

### *Jeneru-tenga*

*Plant characters.* A medium to tall plant, ht. 15-25 ft., upright; densely foliaged; usually thornless, if thorny 2-5 cm. long, one crop annually, heavy fruiter, season January to March.

*Leaf characters.* Leaf medium to large with moderately broad petiolar wing, blade coriaceous, deep green, broadly lanceolate, tendency to curve upwards almost forming a spoon pattern, margin slightly crenate, apex slightly obtuse, emarginate, size 4.5-9.0 cm. long, 2.5-4.5 cm. broad with petiolar wing 2.3-4.8 cm. long, 8 mm.-1.2 cm. broad, big one 10-11.5 cm. long, 5-2.6 cm. broad with petiolar wing 5-5.5 cm. 2.2-2.4 cm.; wing spathulate, margin almost entire, apex obtuse or very slightly truncate; veins slightly prominent on upper side, mix-rib prominent on both sides, articulated.

***Feneru-tenga* (*C. nobilis*)**

*Flower characters.* Bisexual only, usually solitary, occasionally in cluster (2-4), cymes; pedicel, very short, 1-2 mm. long, peduncle 2-6 mm. long, glabrous; calyx medium coriaceous, stellate, apex acute, light ephemeral pink; corolla imbricate, petals 5, ovate-oblong, white (very slightly tinged outside at bud stage, ephemeral), apex almost obtuse or very slightly acute, size 1.4-1.7 cm. long, 6-9 mm. broad; stamens 22-31, filaments free, white, glabrous, 0.7-1.1 cm. long, anther 2 mm. long; ovary oblate, 3 mm. ht., 3-4 mm. diam., 10-13 locules, style terete, white, very short, 5-6 mm. long including stigma, stigma capitate, 2-3 mm. diam.; flowering season March-April.

*External fruit characters.* Shape almost globose to oblate; colour deep orange; surface characteristically warty or pebbled, in certain fruits roughness not very marked, pitted; size usually 5.5-6.4 cm. ht., 6.5-8 cm. diam., big ones 8-8.2 cm. ht., 9-9.5 cm. diam.; wt. 5-8 oz.; D/H index 1.15 to 1.28 (av. 1.18); base obtuse to somwhat flat, basal area wrinkled or rugose; stem short, 0.6-1.3 cm. long; calyx depressed, small, regularly divided, 4-5 lobed, tips both acute and obtuse; apex depressed, navel usually absent, occasionally present, slightly protruded over the apical area, diameter of the navel ring

5-6 mm.; areolar area rough, stylar scar present, small, depressed in an apical pit, style not persistent.

*Internal fruit characters.* Rind medium thick, thickness 5-9 mm. (usually 7 mm.), firmness leathery to medium soft, adherence slight to medium, puffing; oil glands many, conspicuous, number 68-78 per ¼ sq. cm., size and shape variable, primary glands usually oval or ballon-shaped, secondary glands somewhat pyriform, lager than the primary ones; size 1-2 mm. long, 0.5-1 mm. broad, aroma distinct, glandular layer ½ rind; mesocarp (albedo) white, spongy, taste slightly sweet; axis hollow, diameter at the centre 0.8-1.3 cm.; segments 10-13, (usually 12), adherence of carpel medium to slight, septa chartaceous, white, rarely broken at axis, dorsal contour convex to slightly truncate, carpel size 4-5 cm. long, 2-2.4 cm. broad, 1-1.5 cm. thick; pulp-vesicles colour light orange, texture coarse, somewhat loose, not very melting, pulp-wall slightly tough, size 5-9 mm. long, 2-3 mm. broad, stalked, 2-9 mm. long, vesicles arising from dorsal and radial walls of the carpel, shape obovoid to somewhat fusiform, tips both acute and obtuse; juice medium, usually 1½-2½ oz., sweetness and acidity blended, flavour good; seeds few to several, 8-12 per fruit, somewhat cuneate, surface slightly rough, micropydar and usually flat, size 1.4-1.6 cm. long, 6-8 mm. broad 5 mm. thick, testa white, tegmen light brown, chalazal spot brown, cotyledon light green.

The *jeneru-tenga* is fairly common in Upper Assam. The variety is usually grown as a novelty in village homes and has hardly as yet any commercial status. It is a heavy fruiter and the fruits keep longer on the tree than the Khasi orange (*C. reticulata*).

### The Indian Wild Orange (*Citrus indica* Tanaka)

*Origin and name.* Tanaka (1934), states 'It is really a wild citrus found in Nowgong district, Khasi Hills, and Manipur in Assam'. To which it may be further added that *Citrus indica* is found to grow wild in the primeval forests of Rangaphar and Dimapur in the Naga Hills and in Kaziranga Reserve forests in the district of Sibsagar Assam. As a matter of fact, being indigenous to Assam this is well distribtuted throughout the primeval forests of the province.

There is no vernacular name of this citrus and as such the name Indian wild orange, as it has been called, may be retained with advantage to distinguish it from other citrus.

*Nomenclature.* Tanaka (1928) was the first to recognize the Indian wild orange as *C. indica.* Botanical description of *C. indica* as detailed by Tanaka, has been reproduced by Swingle (1946) in the *Citrus Industry* Vol. I., Pp. 419-420. According to Tanaka (1937, a) it belongs to *Meta-citrus* and is much similar to *Citrus depressa* Hayata.

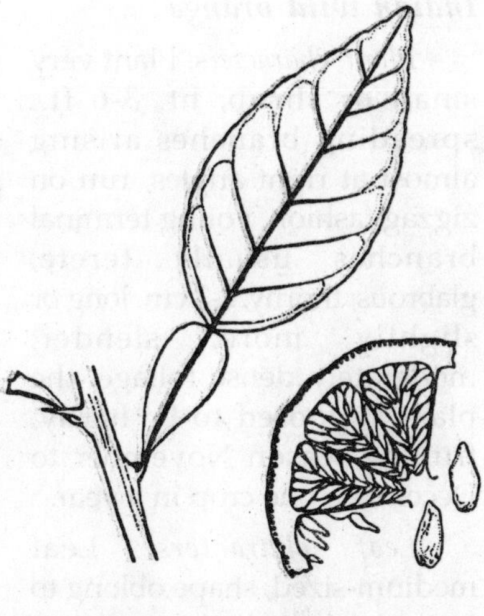

**Jeneru-tenga (C. nobilis) showing outline of leaf and internal structure of fruit.**

In Assam *Citrus Indica* has been collected from different parts of the province and all the specimens tend to show identical morphological characters, which clearly fit in with those of the type species. Apart from the botanical details enumerated by Tanaka, the present writers have been able to add from the study of the fresh sample, a few more details which, it is presumed, will give a more comprehensive picture of this interesting species.

## Key to the Indian Wild Orange (*Citrus indica*)

A. Plant small (ht. 3-6 ft.), branching in zigzag fashion, young branches terete, glabrous spinous; leaf with rudimentary wing or wingless, lanceolate, apex caudate, veins prominent on under surface; flowers solitary, bisexual, stamens 22-25, filaments jointed forming a tube, pubescent.

B. Fruit spherical to depressed globose, rarely sub-pyriform, small, (3-4 cm. diam); surface smooth, deep orange red to scarlet; segments 8-11; pulpvesicles white, polygonal, melting, slimy, sour; seeds few, large, flattened, sub-orbicular, chalazal cap light brown, cotyledon green, monoembryonic.

....................*C. indica.*

### *Indian wild orange*

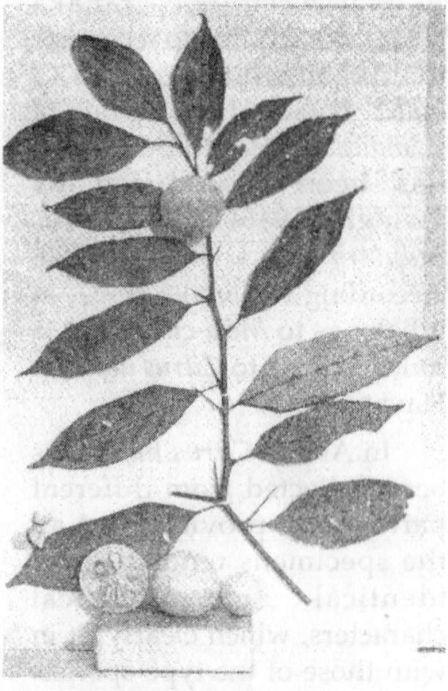

Indian wild orange (*C. indica*)

*Plant characters.* Plant very small or shrub, ht. 3-6 ft., spreading branches arising almost at right angles, run on zigzag fashion, young terminal branches usually terete, glabrous; thorny, 2-3 cm. long or slightly more, slender; moderately dense foliage, the plant supposed to be hardy; fruiting season November to December, one crop in a year.

*Leaf characters.* Leaf medium-sized, shape oblong to lanceolate, gradual tapering at both ends, apex caudate, acute, emarginate, leaf upper surface deep green, under side somewhat lighter, medium coriaceous, margin crenate, also almost serrate towards apex, reticulate venation somewhat prominent on under side, distinct from other common citrus leaves; size 6.5-12.5 cm. long, 2.5-4.5 cm. broad; petiole 0.7-1.5 cm. long, 1-2 mm. broad, petiole short with rudimentary wing articulated.

*Flower characters.* Bisexual, solitary, very rarely in pairs, pedicel very short, 3-4 mm. long, peduncle indistinct, glabrous; calyx greenish white, 4-5 lobed, glabrous, almost cupular, apex deltoid, punctate with convex oil glands on the outer side; coroll imbricate, hypogynous, petals 5, size (in fresh flowers) 1-1.5 cm. long, 3-6 mm. broad, medium coriaceous, petals pure white both outside and inside, apex somewhat obtuse, dotted with convex oil glands on outer side; stamens 22-25, filament white, pubescent, jointed from the base to a considerable length forming a tube, top free, 0.9-1.1 cm. long, anthers introse, 2-3 mm. long; ovary vase shaped, 3 mm. long, 4 mm. broad; fixed on a disk, 8-11 locules, style terete, white, 8-9 mm. long, stigma capitate, 2 mm. diam., flowering season December-February.

*External fruit characters.* Shape spherical to depressed globose and occasionally sub-pyriform; size small ones 2.5-2.8 cm. ht., 2.5-3 cm. dia., normal ones 3.0-3.7 cm. ht., 3.1-4 cm. diam.; wt. ½ oz. to 1 oz. (normal wt. ¾ oz.) : D/H index 1.04-1.14 (average 1.10); colour deep orange red to almost scarlet when full ripe; surface smooth, oil glands usually in flush with the skin, very minutely foveolated; base even, smooth, obtuse; calyx small, stellate, division irregular, short, tip acute, thinly chartaceous; stem 5-7 mm. long; apex obtuse to retuse, areole more or less distinct, in some cases indistinct, areolar area narrow, diam. 0.7-1.4 cm., slightly depressed, stylar scar depressed in a very small apical pit, occasionaly even or minutely raised, style not persistent.

**Indian wild orange (*C. indica*)**

*Internal fruit characters.* Rind very thin, thickness 2 mm., firmness usually soft, adherence slight; oil glands conspicuous, number 18-20 per ¼ sq. cm., primary gland spherical, size slightly less than 1 mm. ht., 1 mm. diam., oil litte, aroma distinct, glandular layer somewhat thick, 3/4 rind; mesocarp (albedo) thin, white, spongy, soft, sweet; axis hollow or semi-hollow, diam at the centre 3-4 mm.; segments 8-11 (usually 10), adherence between carpels slight; rag little; septa chartaceous, white, closed at core, dorsal contour convex, carpel covered with white cotton like substance, carpel size 2-2.4 cm. long, 1.1-1.7 cm., broad, 8 mm.-1 cm. thick; pulp-vesicles white, texture medium coarse, very tender and melting, shape irregular, somewhat fusiform, polygonal, broad and short, arising only from dorsal wall, pulp big ones 5-6 mm. long, 2 mm. broad, stalked, 2-3 mm. long, tip attenuated and obtuse, flavour not agreecable; juice scanty, white slimy (a distinct character), taste very sour; seds few, 6-7 per fruit, very large, size 1.4-1.5 cm. long, 0.9-1.1 cm. broad, 3-5 mm. thick, shape almost round to sub-orbicular with pointed end, flattened, micropylar end acute surface smooth on both sides, testa white, tegmen light brown, chalazal spot light brown, slightly deeper than tegmen colour, cotyledon green, monoembryonic.

The Indian wild orange is never seen under cultivation. It is really a wild citrus found to grow at elevations from 400-5000 ft. or so. The fruits being practically inedible and very small in size do not hold out any promise for commercial exploitation. As a rootstock, it has not been tried in Assam.

## The Tight-skinned Sweet Oranges (*Citrus sinensis* (Linn.) Osbeck)

*Origin and name.* Bonavia (1890) points out that there is no Indian name for the so-called sweet oranges. It is generaly called in India as Maltas indicating that it is not indigenous to India. Swingle (1946) states that the native home of the sweet orange is southern China, Indo-China, or north eastern British India which probably includes some parts of the Naga Hills of Assam. According to Tanaka (1937, a) the sweet orange is suspected to be indigenous to eastern Burma. He also admits of the existence of a closely allied species, represented by *soh-niangling* of Assam. But *soh-niangriang* (Khasi) is the only citrus which actually represents the sweet orange in Assam. The Khasi name *soh-niangriang* has probably been wrongly spelt as *soh-niangling* by Tanaka.

The *soh-niangriang* is met with both in the plains and in the hills of the province and is just grown as a novelty in the gardens. It is reported by Basu (1906) to have originated as chance seedling from the Khasi orange (*C. reticulata*) which is exclusively raised from seeds in Assam. Mitra and Khongwir (1928) suspect it to be either a sport or a hybrid. There seems to be no ground to uphold either of the above presumptions as absolutely identical fruits have been collected from distant Litsimi, a place in the unadministered area towards the north east of the Naga Hills, about 115 miles away from Kohima and bordering almost Upper Burma where no mandarin oranges are found to grow thus ruling out the chance of any inter-crossing. The plants, although reported to be few in number, have been found growing in a wild state far away from human habitation, eliminating thereby the possibility of introduction through human agency. This fruit is called by the Nagas (Sema tribe) as *mishi-thi* which means some kind of sweet orange. The suffix *'thi'* is indicative of Burmese origin. The fruit is identical with the *soh-niangriang* of the Khasi Hills.

Lushai Hill is another place where some kind of sweet oranges almost similar to Maltese Blood red and the like are met with scattered in the villages. It was first thought that these were either introduced by the European Political Officers or by the Christian missionaries of that hill but ultimately it has been ascertained that these were really not introduced fruits. It is claimed by the Lushais to be spontaneously growing there as indigenous to the Lushai Hills. It is also very strange that there is no local name for these varieties although these are claimed to be indigenous. A specimen resembling the Maltese Blood has been collected from Lungleh (Lushai Hills). The fruit though somewhat inferior in quality to the well known Maltese Blood holds out considerabe promise. The variety has not been described in the text for reason of its sporadic occurrence.

In Assam Valley this type of oranges are generally called by the name *mitha-chakla* and in Cachar plains by *jajji-jamir.*

*Nomenclature.* Linnaeus (in *Species Plantarum* 783, 1753) considered the so-called sweet orange or tight-skinned oragne as a variety of the sour orange and called it *Citrus aurantium* var. *sinensis.* Loureiro (1790) designated the sweet orange as *Citrus aurantium,* while Osbeck (1775) earlier segregated it from the sour orange as distinct species and named it as *Citrus sinensis* Linn., which has now been accepted as the correct botanical name for the sweet oranges. The distinctive characters that led to raise the sweet orange to a specific rank have been discussed by Swingle (1946).

**Key to variety of the sweet orange**

A. Tree medium to tall, erect, spreading, deep green foliage, thorny, leaves broad lanceolate with very narrow petiolar wing.

B. Fruit usually globular, size medium (7-7.5 cm diam.); surface smooth, colour bright yellow to orange; rind adherence strong with carpels; juicy, sweetness and acidity blended; seed somewhat cuneate to obovate, chalazal spot Indian red.

### *Soh-niangriang*

**Soh-niangriang (*C. sinensis*)**

*Plant characters.* A medium to tall plant, ht. 20-25 ft., trunk straight and clean, upright, spreading branches, densely foliaged; thorny, spines slender, big ones 2-6 cm. long; fruiting season December to January, prolific fruiter, one crop annually.

*Leaf characters.* Medium-sized, shape somewhat broadly lanceolate medium coriaceous, deep green, margin crenate particularly confined to the top half of the blade, apex attenuated, acute, occasionally emarginate; petiole (when narrowly winged) spathulate, margin entire, apex obtuse articulated, blade size 6.5-10.5 cm. long, 3.0-5.2 cm. broad with petiolar wing 1-2 cm. long, 1-3 mm. broad.

*Flower characters.* Both solitary and glomerate (1-8), bisexual only, cymes; pedicel shrot, 5-6 mm. long, peduncle 0.5-1.1 cm. long, glabrous; calyx medium thick, somewhat cupular, greenish white, 4-5 lobed; corolla imbricate, petal 4-5, pure white. oblong, apex slightly acute, size 1.7-2.1 cm. long, 5-9 mm. broad; stamens 20-30, filaments polyadelphous, 1 or 2 free, white, glabrous, 1-1.5 cm. long, anther 2-3 mm. long; ovary globose, size 2-3 mm. long, 2-3 mm. broad, style white, cylindrical, 0.8-1 cm. long, stigma capitate 2-3 mm. diam; flowering season March-April.

*External fruit characters.* Shape usually globular to almost spherical, occasionally slightly turbinate; size 6.5-7.2 cm. ht., 7-7.6 cm. diam. wt. 6-8 oz.; D/H index 1.04-1.17 (av. 1.08); colour bright yellow to orange; surface smooth, very finely pitted, somewhat glossy; base obtuse, basal area generally smooth, occasionally very minutely ribbed; calyx even or occasionally depressed, medium coriaceous, irregularly divided, division short, deltoid; stem small, 1 cm. long; apex obtuse to truncate, areole absent, areolar area smooth, stylar scar present, small even or depressed in an apical pit, style not persistent.

*Internal fruit characters.* Rind medium thick, thicknes 4-7 mm. (usually 5 mm.), firmness leathery to medium soft, adherence strong; oil glands conspicuous, number 26-32 per ¼ sq. cm., shape usualy pyriform, also spherical or ellipsoid, size 1-3 mm. long, 0.5-2 mm. broad, oil abundant, aroma distinct, glandular layer about ½ rind : mesocarp (albedo) white, spongy, very slightly weet or rather tasteless; axis usually solid; segments 10-13 (usually 11), septa white, tough, chartaceous, dorsal contour convex to slightly turncate; rag little; pulp-vesicles honey yellow to light orange, texture coarse, melting, shape irregular, usually fusiform,

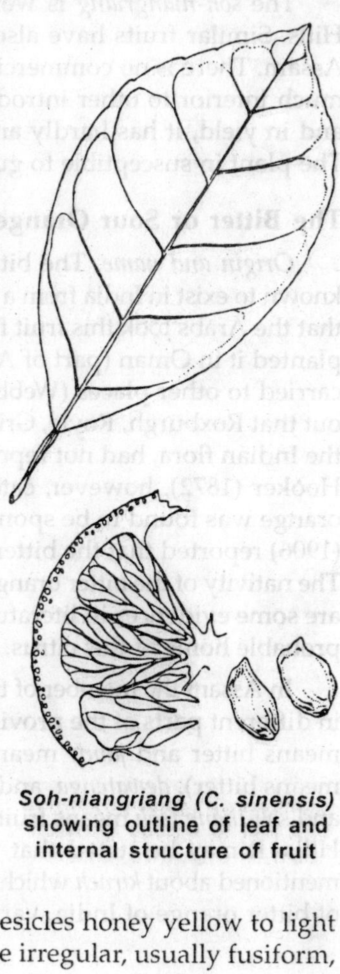

***Soh-niangriang (C. sinensis)* showing outline of leaf and internal structure of fruit**

tips acute and obtuse, stalked, pulp size big ones 0.9-1.3 cm. long, 2-4 mm. wide, vesicles arising from dorsal and lower half of the carpellary walls, flavour goods; juice abundant, 3-4 oz. per fruit, colour light orange, sweetness and acidity blended in full ripe fruits, seeds several, 15-25 per fruit, size 1.3-1.6 cm. long. 0.8-1.1 cm. broad, 5-7 mm. thick, shape cuneate to obovate, micropylar end somewhat beaked, slightly ridged on one side, other side with some smooth area, testa cream white, tegmen light brown, chalazal cap Indian red (slightly deeper in colour than that of the *Valencia*), cotyledon white.

The *soh-niangriang* is well distributed throughout the Khasi Hills. Similar fruits have also been obtained from other parts of Assam. There is no commercial use of this fruit as yet. Being very much inferior to other introduced sweet oranges both in quality and in yield, it has hardly any commercial possibility in Assam. The plant in susceptible to gummosis.

## The Bitter or Sour Oranges (*Citrus aurantium* Linnaeus)

*Origin and name.* The bitter or sour orange as it is called, is known to exist in India from a very ancient past. There is a reference that the Arabs took this fruit from India in about 922 A.D. and first planted it in Oman (part of Arabia) and subsequantly it had been carried to other places (Webber, 1946). Tolkowsky (1938) pointed out that Roxburgh, Royle, Griffith and Wight, the first explorers of the Indian flora, had not reported about its existence in India; but Hooker (1872), however, categorically mentioned that the bitter orange was found to be spontaneously growing in India. Brandis (1906) reported that the bitter orange is rarely cultivated in India. The nativity of the bitter orange is shrouded with mystery. But there are some evidences in literature to justifiably claim India to be the probable home of this citrus.

In Assam the member of this group goes by various local names in different parts of the province. It is known as *karun-jamir* (*karun* means bitter and *jamir* means Citrus) in Cachar, *tita-karuna* (*tita* means bitter), *dewatenga*, and *chakla-tenga* in Assam Valley plains, and *soh-thang* (*soh* means fruit, and *thang* means bitter) in the Khasi Hills. It may be stated that Babar (1519) in his famous memoir mentioned about *kirneh* which Bonavia (1890) considers to be a form of bitter orange of India, variously called *khatta* (sour) and *Karna* (bitter). The general Indian name *'karna'* has a very striking phonetic similarity with the *karun* or *karuna* of Assam. It is not improbable that *karun* or *karuna* owes its origin from the common Indian word *kirneh* or *karna* referred to above.

In Assam the variety is often seen growing almost gregariously in a semi-wild state, particularly in certain parts of the Naga Hills. It is regularly cultivated in the Khasi Hills and in Cachar, though its uses are very much restricted. Robinson (1841) indicated the existence of *tita-karuna* (bitter orange) in this province. The *chalkla-tenga* or *sakla-tenga* which is a form of the bitter orange had also been referred to by Anderson of Sibsagar Assam, while furnishing a list of Citrus varieties growing in Assam to Bonavia (1890). There is no reliable evidence, however, to show whether the sour or bitter orange in an introduced fruit or is indigenous to Assam. That it is well distributed throughout the province is more than certain, including that even if it be an introduced fruit it must have been acclimatized from a very distant past in Assam.

*Nomenclature.* The sour or bitter orange has been differently named by different authors, of which *Citrus aurantium* Linn. (1753), *Citrus vulgaris* Risso (1813), *C. bigaradia* Risso et Peiteau (1818-1822), *C. amara* Linn. (1831), *C. karna* Raf. (1838), *C. aurantium* var, *bigaradia* Hooker (1872), have been in frequent use. In old classification *Citrus aurantium* Linn. is used to represent all oranges in general including sour or bitter orange. But it has now been agreed to by all taxonomists that this name should exclusively stand for the bitter orange. The difference between the bitter orange and the so called sweet orange (*C. sinensis*) is obvious and distinct. Their morphological and other differences have been fully discussed by Swingle in the *Citrus Industry,* Vol. I., Pp. 402-406, which may be read with interest.

Webber (1946) divides the sub-species *amara* (Seville group) into two distinct groups namely (i) normal bitter orange group and (ii) aberrant group. The broad distinction between the two has also been discussed by him. Besides, two other sub-groups, viz. (i) Myrtifolia and (ii) Bitter sweet have also been recognized; the origin of which has been attributed by him to mutation or sport.

In the light of Webber's division the bitter orange of Assam has been segregated into (i) normal and (ii) aberrant group.

**Key to varieties of the bitter orange**

A. Normal Group.

B. Tree tall, erect, taller than broad; upright branching; densely foliaged, dark green, glossy; leaf broadly lanceolate; petiole broadly winged, obcordate; leaf aroma characteristically typical of the sour orange, assisting diagnosis.

C. Fruit globose to obovoid; size medium (6.4-8.0 cm. diam.); surface almost smooth to rugose, colour deep orange to almost scarlet, mesocarp white, bitter; juice very sour with soupcon of the bitterness; colour light orange to antimony yellow; seeds cuneate, chalazal spot Indian red.

$V_1$.....*Karun-jamir*

AA. Aberrant group.

B. Tree medium-sized, erect; very much spreading branches, broad round top; densly foliaged, deep green, leaf broadly different from that narrowly winged, spathulate, leaf aroma slightly different from that of the normal group.

C. Fruit globular to slightly tribunate; size larger than the typical sour orange, (9.0-10.8 cm. diam.); surface smooth, yellow to golden yellow; mesocarp white, bitter; very juicy, intensely sour, colour white; seeds few, shape ovate to cuneate, chalazal spot Indian red.

$V_2$......................*Gondh-huntra.*

***Karun-jamir***

*Plant characters.* A medium to talt plant, 25-30 ft., erect; branches not spreading usually taller than broad; very densely foliaged; usually thornless; fruiting season December to March, one crop annually, medium fruiter.

Karun-jamir (*C. aurastium*)

*Leaf characters.* Leaf medium to large with broad petiolar wing; size 8.5-15 cm. long, 4.0-7.6 cm. broad with petiolar wing 2.0-4.5 cm. long, 2.5 cm.

broad; shape (blade) broadly lanceolate, medium coriaceous, dark green, articulated, margin crenate, apex attenuated and acute, occasionally emerginate; petiolar wing usually obcordate, occassionally spathulate, apex retuse in broad wing, rounded in spathulate wing; crushed leaf aroma typical.

*Flower characters.* Flower bisexual only, both solitary and glomerate (1-4), cymes; pedicel 4-6 mm. long, peduncle 0.4-1 cm. long, glabrous; calyx medium coriaceous, greenish white, deltoid, apex acute, ciliate, 4-5 lobed; corolla imbricate, petals 4-5 pure white, medium thick, oblong, apex attenuated and acute, size 2-2.5 cm. long, 5 mm. -1 cm. broad; stamens 22-27 (usualy 22-24), filament white, polyadelphous, 1 or 2 free, 1.4-1.8 cm. long, glabrous, anthers usually 2-3 mm. long, ovary 4 mm. long, 3 mm. broad, style cylindrical, white, 1.2-1.3 cm. long including stigma, stigma capitate, flowering season March-April.

*External fruit characters.* Shape globose to obovoid, size 6.4-8.5 cm. ht., 6.4-8.2 cm. diam.; weight 5½-10 oz. (usually 7-9 oz.); D/H index 0.96-1.15 (av. 1.04); colour deep orange to scarlet; surface smooth to slightly rugose, finely pitted, glossy; base obtuse, occasionally slightly ribbed; calyx small, irregularly divided; stem 1-1.5 cm. long; apex usually even, obtuse, rearely slightly depressed, navel absent; areole inconspicuous, stylar scar present, either even, or slightly depressed in a small apical pit, style not persistent.

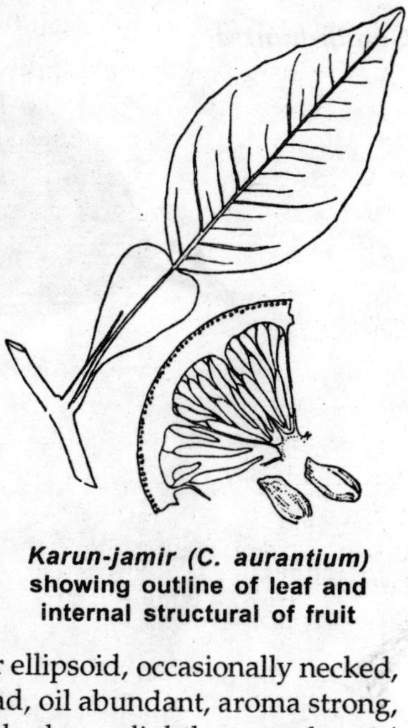

**Karun-jamir (*C. aurantium*) showing outline of leaf and internal structural of fruit**

*Internal fruit characters.* Rind medium thick, thickness 0.7-1 cm., usually 7-8 mm., firmness medium soft, adherence strong; oil glands conspicuous, number many, 30-50 per ¼ sq. cm., shape usually oval or ellipsoid, occasionally necked, size 0.5-2 mm. long, 0.5-1 mm. broad, oil abundant, aroma strong, typical of true sour orange, glandular layer slightly more than 1/3 rind; mesocarp (albedo) white, soft and spongy, taste bitter; axis holow or semi-hollow, mid diameter 1-1.3 cm.; segments 9-11, adherence somewhat strong, rag soft and little; septa medium coriaceous, white, slightly tough, dorsal contour convex to truncate, carpel size usually 4.8-6.5 cm. long, 2.2-2.8 cm. broad, 1-2.5 cm. thick; pulp-vesicles melting, arising only from dorsal side, shape irregular, generally fusiform, size big ones 1.2-2.2 cm. long, 2-3 mm. broad, stalked, tip usually slightly acute to obtuse; juice abundant, weight 2½-5 oz., colour light orange or antimony yellow, taste very sour with soupcon of bitterness; seeds many 30-50 per fruit, size 1.2-1.6 cm. long, 7-9 mm. broad, 4.5 mm. thick, somewhat cuneate, micropylar end usually flat, one side usually with a smooth area and other with ridges, testa cream white, tegmen light brown, chalazal cap corinthian red (Indian red), cotyledon white.

The *karun-jamir* is not grown as a commercial crop in Assam. The small production is disposed of in the local markets for meeting ordinary local needs. It is supposed to be long-lived and can withstand extreme conditions of soils and climates. As a root-stock this variety does not appear to be compatible with the Khasi orange (*C. reticulata*).

### *Gondh-huntra*

**Gondh-Huntra (*C. aurantium*)**

*Plant characters.* Plant medium-sized, ht. 22-22 ft., erect; very spreading branches, densely foliaged; thorny, 0.5-4 cm. long, slender; young parts angular, glabrous; fruiting season November to February, one crop annually, medium fruiter.

*Leaf characters.* leaf medium to large; shape (blade) broadly lanceolate, deep green, apex attenuated, acute, emarginate, margin, crenate, petiole slightly obcordate to spathulate, apex obtuse to sligthly retuse, margin almost entire to slightly crenate, articulated; size (blade) 6-14 cm. long, 3.2-7.5 cm. broad, with petiolar wing 1.5-3.4 cm. long, 4 mm. -1.6 cm. broad, crushed leaf aroma distinct.

*Flower characters.* Flowers solitary and clustered (1-7), (usually 1-3), bisexual only, cymes; pedicel 4-6 mm. long, peduncle 5 mm.-4 cm. long, glabrous; calyx medium coriaceous, greenish white, deltoid, apex acute, 4-5 lobed; corolla imbricate, petals 4-5, white, oblong, size 1.9-2.2 cm. long, 8 mm.-1.1 cm. broad; stamens 20-33 (usually 24-27), filaments polyadelphous, some free, white, glabrous, 1-1.5 cm. long, anther 2-3 mm. long; ovary oval-shaped, 3-4 mm. long, 3 mm. broad, 8-11 locules, style cylindrica, white 1-1.1 cm. long including stigma, stigma capitate; flowering season March-April.

*External fruit characters.* Shape usually globular to slightly turbinate, size 9-10.5 cm. long, 9.1-10.8 cm. diam.; weight 12-20 oz. (usually 14-18 oz.); D/H index 1.00 to 1.11 (average 1.07); colour yellow to golden yellow; surface smooth, glossy, very finely pitted; base almost obtuse, very slightly narrowed, smooth, occasionally slightly ribbed; apex obtuse, slightly depressed, areolar mark indistinct, stylar scar present, depressed in a small apical pit, style not persistent.

*Internal fruit characters.* Rind medium thick, thickness 0.6-1 cm., firmness medium soft, adherence medium; oil glands conspicuous; number 20-24 per ¼ sq. cm., shape variable, spherical,

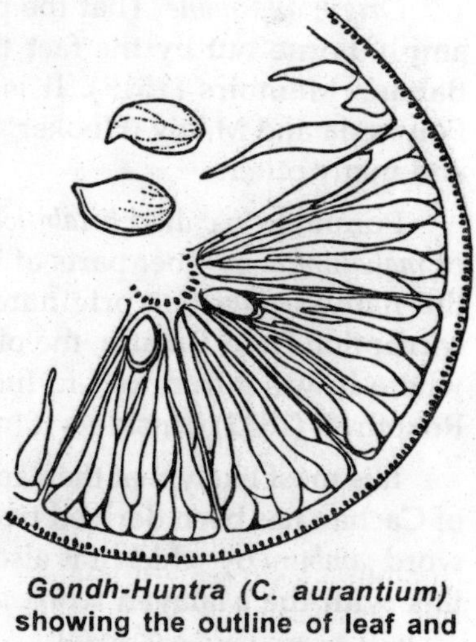

**Gondh-Huntra (*C. aurantium*) showing the outline of leaf and internal structure of fruit**

ovate, or pyriform, size 0.5-4 mm. long, 0.5-2 mm. broad, oil abundant, glandular layer obout ½ rind; mesocarp (albedo) white, spongy, soft , bitter, vascular bundle prominent; axis hollow basal diameter 0.7-1.0 cm., mid diam. 1.5-2 cm., apical diam. 0.8-1 cm., segments 8-11, adherence medium strong; rag very soft, little; septa chartaceous, white, slightly tough, dorsal contour convex to slightly truncate, capel size 6.5-8 cm. long, 3-3.5 cm. broad, 1.5-3.5 cm. thick, carpel occasionally broken at axis; pulp-vesicles white, texture coarse, very melting, arising from dorsal side, fusiform. tips usually obtuse, size big ones 1.5-2.7 cm. long, 3-5 mm. broad, stalked; juice abundant, 7-12 oz. per fruit, (usually 8-9 oz.) colour white, very sour, not bitter, flavour good; seeds several, 2-32 (usually 10-15) per fruit, size 1.4-1.7 cm. long, 0.8-1 cm. broad, 5-7 mm. thick, shape slightly ovate to cuneate, plano-convex or double convex, beaked or flat at micropylar end, one side smooth other side marked with anastomosing lines, rounded end somewhat rough and wrinkled, testa cream white, tegmen light brown, chalazal cap Indian red, cotyledon white.

The plant grows very vigorously. The fruits are very juicy and used generally for *sherbet* (cold drink) and for cleaning brass utensils. It is sold in the markets as a minor fruit having no commercial importance at present.

### The Pummelos (*Citrus grandis* (Linn.) Osbeck)

*Origin and name.* That the pummelo is not of Indian origin is amply borne out by the fact that it has not been mentioned in Babar's Memoirs (1519). It is considered to be indigenous to Polynesia and Malay (Hooker, 1872, Brandis, 1874, Webber, 1946) and many others.

Pummelo is called *batabi-lebu* in Bengal and *chakotra, mahatabi* or *mahanimbu* in other parts of India. Bonavia (1890) assumes that the name *batabi* has originated from Batavia and *chakotra* by corruption from Fakatra, the old name of the capital of Java from where it was introduced to India in comparatively recent times. Roxburgh (1832) considers it to be of exotic origin.

It is most likely that the name *matu,* as it is called in the district of Cachar, has been derived from the word *mahatabi.* Similarly the word *jambura* by which it is also known in Sylhet might have some link with the Malayan word *jamboa* for the pummelo. All these

Indian names as enumerated above clearly indicate the possibilities that the pummelo has been introduced into India from a common source. The common Assamese word for pummelo is *rabab-tenga.* To get at the derivation of this word nothing can be said beyond what McLeod infers (refer Plate No. CCXGXVIII, Bonavia). He says that it is possibly a corruption of the word *Nawab,* the king, an attribute perhaps to the big size of the fruit. But his assumption seems to be based on far-fetched imagination. In the Lower Assam Valley the pummelo is also occasionally known as *hukma-tenga* and in the Khasi Hills as *soh-myngor* or occasionally *soh-saw.* Tanaka (1937, a) reported that Brandis collected Shaddock or pummelo wild in Darmaghat (Garo Hills). It seems to be doubtful, because it is most likely that Brandis collected his sample from some interior village abandoned by the Garos who like all other hill tribes of Assam are also nomadic in their nature and thus Brandis might have inferred that it was growing wild.

Nothing definitely can be said with regard to introduction of the pummelo into Assam. It is well distributed both in the plains and in the hills to an altitude of 5000 ft. or so. The trees are usually heavy yielders. The fruit is cheaper than the other citrus and is valued for its medicinal properties.

*Nomenclature.* The taxonomic difference between the pummelo and the grapefruit is very little and the distinguishing characters that separate one from the other may just be reckoned with mere varietal manifestations and not specific. Besides, grapefruit is considered by many to have originated from the pummelo as a mutation (Tanaka, 1932; Hume, 1941; Webber, 1946), or as a hybrid of the pummelo with the sweet orange (Swingle, 1946). It is, therefore, evident that the separate specific recognition of grapefruit (C. *paradisi*) is a matter of only long standing usage in commerce but has hardly any taxonomie justification. Macfadyen (1830 and 1837) separated the grapefruit from the pummelo or shaddock and named it as C. *paradisi* as distinct from the pummelo (C. *grandis*).

The pummelo has been called by various botanical names of which C. *grandis* (Linn.) Osbeck C. *decumana* Linn., C. *maxima* (Burm.) Merril occur frequently in literature. In old publications on Indian flora the pummelo has been referred to as C. *decumana* (Roxburgh, 1832; Brandis, 1874; Cooke, 1903; Lushington, 1910). Recently Swingle (1946) considers it appropriate to retain the oldest valid name C. *grandis* (Linn.) Osbeck for this fruit. Tanaka (1937, a) also favours retaining the same. Chatterjee (1948), however, points out that the correct name of the pummelo should be C. *maxima* (Burm.) Merri) on the grounds of priority of publication. In view of this nomenclatural controversy the writers of not propose to enter into any discussion on the issue in their monograph. They leave this to the competent authority to decide upon the validity of rival claims. The present authors, however, deem it expedient to follow Swingle and Tanaka in this case to minimize confusion.

There are many forms of the pummelo developed naturally in Assam. All these can broadly be reduced to four different categories basing on (a) shape, (b) size, (c) colour of mesocarp, and (d) pulp-vesicles. The different specimens so far collected and studied in Assam have been sifted out into nine distinct forms based on the combinations of the different distinguishing characters referred to above. It is not unlikely that some more may still be there and which have escaped our notice since the range of variation is wide in the pummelo. In Bonavia's Plate No. CCXXVIII a form of the pummelo of Assam has been represented, and in his other plates LXX to LXXXVIII many other varieties have been depicted showing the range of diversity of Indian forms.

All the forms in Assam go usually by these common group names, eg. *rabab-tenga, matu,* and *soh-myngor* in different places irrespective of size, shape, colour of albedo and pulp, etc. and it would be a misnomer to try to affix a varietal name to each and every form. It should be realised that none of the forms of the pummelo has any particular importance as a horticultural variety and any attempt at separating one from other through vernacular group names would be only leading to confusion. It is, therefore, proposed to deal with the different forms occurring within this group in a general way, calling the variety after the name of the place from where it has been collected.

**Key to varieties of the Pummelo**

A. Tree medium to very tall; very much spreading, twings usually pubescent; leaves large with broad petiolar wing; shape ovate-oblong or broadly lanceolate; petiolar wing prominent, also narrowly winged, shape cordate to stipulate; mid-rib often pubescent; flowers very large, usually clustered, raceme, pedicels and peduncles pubescent, petals 4-5, pure white, stamens many.

B. Fruit large to very large, mainly borne singly; axis usually hollow or semi-hollow; pulp loose, not melting, acidulously sweet, bitter principle more or less marked, pulp arranged in herringbone fashion; seeds large, usually cuneate, rough, chalazal cap light brown to deep brown.

C. Shape turbinate; pulp light pink.

$V_1$.............*Dowadi.*

CC. Shape spherical; pulp light pink.

$V_2$.............*Nowgong.*

CCC. Shape ellipsoid; pink fleshed

$V_3$.............*Jorhat.*

CCCC. Shape oblate; pink fleshed.

$V_4$.............*Khananara.*

CCCCC. Shape ovate-pyriform; pulp rose coloured.

$V_5$.............*Kamrup.*

CCCCCC. Shape spherical; pulp venison or beef coloured

$V_6$.............*Khasi*

CCCCCCC. Shape spherical; pulp white.

$V_7$.............*Burni.*

BB. Fruit medium sized (size of standard grapefruits); pulp melting, bitterness not marked; chalazal spot sepia brown.

C. Shape turbinate; pulp colour honey yellow

$V_8$.............*Zemabawk.*

CC. Shape slight roundish to oblate; pulp colour light yellowish white.

$V_9$.............*Aijal.*

***Dowadi (light pink flesh)***

*Plant characters.* A tall plant, ht. 20-30 ft., very much spreading branches, densely foliaged; almost thornless; young shoots pubescent, angular; tree supposed to be hardy and long-lived; fruiting season October to December, one crop in a year.

*Leaf characters.* Leaf large with broad wing, coriaceous, dark green, blade size 10-13 cm. long, 5.3-8.0 cm. broad; shape ovate-oblong, apex attenuated and slightly obtuse, emarginate, margin crenate or bi-crenate; vein very slightly prominent on both sides, mid-rib pubescent on the lower side; petiole broadly winged, cordate, size 4-6.5 cm. long, 2.6-4.8 cm. broad, apex retuse, margin crenate, articulated.

*Flower characters.* Bisexual only, glomerate (4-10), large, raceme; pedicel 1-1.8 cm. long, peduncle 0.6-2.2 cm. long, pubescent; calyx tube coriaceous almost cupular, irregularly divided, apex either acute or obtuse, white; corolla imbricate, petals 4-5, pure white oblong, coriaceous, dotted with oil glands outside, tips almost obtuse, size 2.1-2.8 cm. long, 0.8-1.3 cm. broad; stamens 31-36, anthers 3-4.5 mm. long, filament polyadelphous, some free, 1-1.5 cm. long, glabrous, white; style cylindrical, white, glabrous 1.1-1.2 cm. long including stigma, stigma capitate, large, 3-4 mm. diam.; ovary usually 4 mm. long. 3 mm. broad, flowering season March-April.

*External fruit characters.* Shape turbinate to ovate-pyriform, size large, 13-14.5 cm. ht., 12.4-13.8 cm. diam., occasionally slightly bigger; wt. 1½-3 lb., (usually 2 lb.); D/H index 0.93-0.95 (av. 0.94); colour light yellow when ripe; surface smooth, papillate; base moderately depressed, very low-necked, basal area (stalk mark) 7-9 mm. diam., slightly ribbed; stem large, 2.5-3 cm. long; calyx large, irregularly divided, tips obtuse, reflexed upwards, medium coriaceous; apex obtuse depressed, areole absent, surface smooth, occasionally very sligtly rugose, stylar scar depressed in small apical pit, style not persistent.

*Internal fruit characters.* Rind thick, thickness 1.5-2 cm. (in median cross section), firmness brittle, adherence medium strong; oil glands conspicuous, number 12-14 per 1/4 sq. cm., primary gland spherical to ellipsoid, size 1-2.5 mm. long, 1-2 mm. broad, surface convex, aroma distinctly of pummelo, glandular layer about 1/5 of rind; mesocarp (albedo) light pink, deeper in inner side, vascular bundle prominent texture soft and spongy taste slightly sweet with soupcon of bitterness; axis usually hollow, irregular in outline in median cross section, mid diam. 1.2-1.6 cm., segments 10-13, adherence medium strong; rag coarse, somewhat hard, medium; septa thick, leathery, tough, both broken or closed at axis, colour deep pink, dorsal contour convex to turncate, occasionally slightly retuse; pulp-vesicles colour not uniform, tips very light pink, base pink, loose, separable without breaking the pulp, arranged in herringbone fashion, arising from dorsal and from lower portion of the carpellary walls, vesicles irregular in shape and size, tips obtuse, size big ones 2-2.2 cm. long, 4-5 mm. broad, small ones 0.6-1.2 cm. long, 4-5 mm. broad, coarse grained, not very melting, flavour good; juice abundant, pinkish, wt. 7-10 oz., taste sweet and acidulous, bitter principle mild; seeds 30-50 or more, shape usually cuneate with flat micropylar end, one side smooth, plano-convex anastomosing lines on rounded end, size usually 1.7-2 cm. long, 1 cm.-1.3 cm. broad, 5-6 mm. thick, testa white or very slight cream white, tegmen light pinkish brown, chalazal spot Indian red, cotyledon white, monoembryonic.

## *Nowgong (light pink flesh)*

*Plant characters.* A tall plant, 25-30 ft., erect; well branched, very much spreading, densely foliaged; almost thornless; young shoot pubescent; supposed to be long-lived; fruiting season early November to December, one crop in a year.

*Leaf characters.* Leaf large, broadly winged, coriaceous, deep green, blade ovate oblong, margin almost entire or slightly crenate, apex almost obtuse, size 12-15.5 cm. long, 6-8.5 cm. broad with petiolar wing 3.5-6 cm. long, 1-4.5 cm. broad; petiole winged, shape obcordate, apex truncate to retuse, margin entire; mid-rib pubescent on the lower side; articulated.

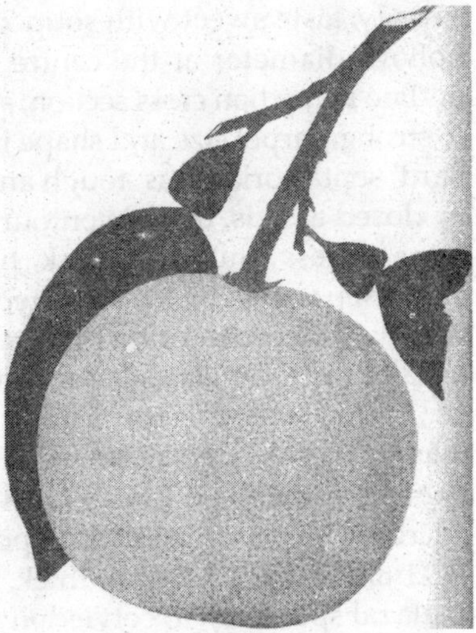

**Nowgong-*rabab* (*C. grandis*)**

*Flower characters.* Flower large, bisexual only, glomerate (6-8), raceme; pedicel 1.5-2 cm. long, peduncle 1.5-2.5 cm. long, pubescent; calyx coriaceous, cupular, division irregular, apex both acute and obtuse, 4-5 lobed; corolla imbricate, definitely 4-merous, thick oblong, apex almost obtuse, size 3-3.2 cm. long, 1.2-1.5 cm. broad, petals pure white, reflexed; stamens 32-38, anther 5-6 mm., style terete, white, 1.1-1.2 cm. long including stigma, stigma diameter 5-6 mm.; ovary spherical usually 5 mm. long, 5-6 mm. board, 18-22 locules, flowering season February to April.

*External fruit characters.* Shape almost spherical to globose; size large, 13.2-15.5 cm. ht., 14.4-16.6 cm. diam., wt. 2 lb. to 4 lb.; D/H index 1.04-1.10 (av. 1.08); colour light yellow; surface finely papillate, oil glands conspicuous; base evenly obtuse, occasionally slightly ribbed, stem somewhat large, 2-2.8 cm. long; calyx large, irregularly divided, tips acute or obtuse, persistent reflexed; apex moderately depressed, smooth; areole absent, stylar scar depressed in a small apical pit, style not persistent.

*Internal fruit characters.* Rind thick, thickness 1.5-1.8 cm. (in median cross section), firmness firm and brittle, adherence medium to somewhat strong; oil glands convex, conspicuous, numbr 14-16 per 1/4 sq. cm., primary glands somewhat spherical to oval, size 1.5-2 mm. ht., 1-2 mm. diam., aroma distinctly of pummelo, glandular layer about 1/5 rind; mesocarp white, texture soft and spongy, taste sweet with soupcon of bitterness; axis usually semi-hollow, diameter at the centre 1.5-3 cm., irregular or oblique in outline in median cross section; segments 18-22, adherence medium to strong, carpel size and shape irregular; rag coarse and somewhat hard; septa coriaceous, tough and leathery, light pink, both broken or closed at axis, dorsal contour convex to truncate; pulp-vesicles more or less uniformly pink, tips slightly whtie, texture coarse, easily separable, loose, arranged like herringbone fashion, arising usually from dorsal side of the carpel, irregular in shape and size, usually fusiform, flavour good; juice abundant, light pink, wt. 8-13 oz., taste sweetness and acidity blended, bitter principle slightly marked; seeds large, many, 60-80 per fruit, shape irregular, usually cuneate, one side plano-convex, other ridges and furrows, micropylar end either slightly pointed or flat, size 1.6-1.8 cm. long, 1-1.1 cm. broad, 5-6 mm. thick, testa white, tegmen light brown, chalazal spot brown, cotyledon white.

## *Jorhat (Pink flesh)*

*Plant characters* A tall tree, ht. 25-30 ft., erect; well branched; densely foliaged; almost thornelss; young parts pubescent; fruiting season November to January, one crop in a year.

*Leaf characters.* Leaf large with broad wing, coriaceous, deep green, margin crenate, apex attenuated and obtuse, emarginate, shape ovate-oblong, size 9.5-12.1 cm. long, 6-7 cm. broad, petiole broadly winged, size 3-4 cm. long, 1.5-3.5 cm. broad, shape obcordate, apex truncate to slightly retuse; veins very slightly prominent on both sides, mid-rib pubescent on lower surface.

*Flower characters.* Flower large, bisexual only, glomerate (3-10), raceme, pedicel 1.3-1.8 cm. long, peduncle 1.5-5.5 cm. long, pubescent; calyx coriaceous, cupular, division deep, apex obtuse or slightly acute, usually 4-lobed; corolla imbricate, petals 4-5, pure white oblong, apex obtuse, 2.3-3.1 cm. long, 0.9-1.4 cm. broad, stamens 34-41, anther 3-5 mm. long, filaments polyadelphous, some free, glabrous white, 1.2-1.8 cm. long; ovary shape usually like bottle neck, size 8-9 mm. lon,g 5-6 mm. broad, style cylindrical, white, 1-1.1 cm. long including stigma, stigma capitate, diameter 4-5 mm., flowering season February to April.

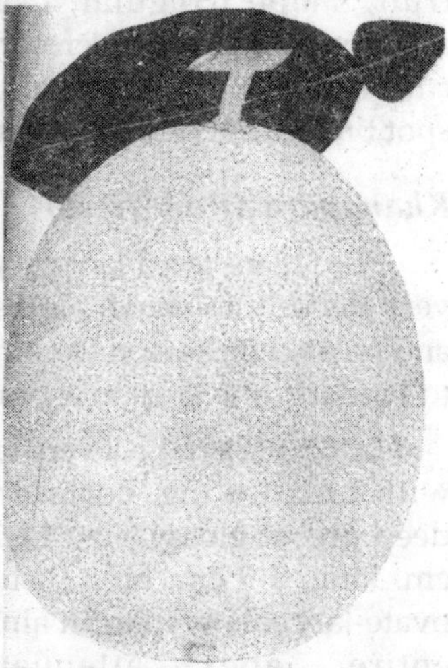

**Jorhat-rabab (*C. grandis*)**

*External fruit characters.* Shape ellipsoid, large, variable, size 17-18.8 cm. ht., 14.5-15.6 cm., diam, wt. 2½-4 lb.; D/H index 0.80-0.85, (av. 0.82); colour light yellow, occasionally with a dash of pink; surface smooth, papillate, very minutely pubescent; base evenly rounded, occasionally ribbed; stem 1.52 cm. long; calyx large, coriaceous, 3-4 lobed, tips reflexed; apex moderately depressed, smooth; areole absent, stylar scar depressed in a small apical pit, style not persistent.

*Internal fruit characters.* Rind thick, thickness 1.0-2.1 cm. in median cross section, firmness firm to brittle, adherence medium to strong; oil glands large, spherical to ellipsoid, 2 mm. ht., 2 mm. diam., number 10-12 per ¼ sq. cm., surface convex, aroma distinctly of pummelo, glandular layer about 1/5 rind; mesocarp (albedo) pink, spongy, bitter with soupcon of sweetness, axis hollow, mid. diam. 1-2 cm., irregular in outline; segment 12-14, adherence medium, carpel size and shape irregular, carpel size 13 cm. long, 4.5 cm. broad, 2-2.4 cm. thick, septa thick, tough, both broken and closed at axis, colour pink, dorsal contour turncate to slightly retuse; pulp-vesicles uniformly pink, tips somewhat colourless, texture coarse, easily separable, loose, arranged in herring-bone fashion, arising mostly from dorsal side of carpel, usually fusiform, tips slightly pointed and blunt, not melting, flavour good; juice plenty, colour pink, wt. 12-15 oz., taste acidulously sweet, bitter principle not very marked; seeds many, large, numbr 100-130 per fruit, shape irregular, usually cuneate, plano-convex with anastomosing lines, micropylar end flat, size 2-2.2 cm long, 1.1-1.2 cm. broad, 5-6 mm., thick, testa white, tegment light brown, chalazal spot brown, cotylendon white.

### *Khanapara (Pink flesh)*

*Plant characters.* Plant very tall, ht. 30-40 ft., erect; well branched; very densely foliaged, almost thornless; young branches angular and pubescent; season December to January, one crop in a year.

*Leaf characters.* Leaf very large with broad wing, coriaceous, deep green, glossy, size 12-16.8 cm. long, 5-9 cm. broad, shape ovate-lanceolate, margin almost entire, apex attenuated, emarginate; petiole wigned, size 2.5-4.8 cm. long, 1.5-3 cm. broad, shape obcordate, apex truncate to retuse, margin entire; mid-rib pubescent on underside, articulated.

**Khanapara rabab (*C. grandis*)**

*Flower characters.* Flower very large, bisexual, glomerate (2-12), raceme; pedicel 2-2.3 cm. long, peduncle 3-10 cm. long, pubescent; calyx large, coriaceous, cup-shaped, tips deltoid, 4-5 lobed; corolla imbricate, petals usually 4 (rarely 6), pure white, oblong, apex almost obtue, size 3-3.4 cm. long, 1-1.5 cm. broad; stamens 40-42, filament white, polyadelphous, and some free, (triangular in cross section); glabrous, 1.5-1.9 cm. long, anther 4-6 mm. long, ovary 6-7 mm. long, 6-8 mm. broad, style white, cylindrical, 1.4-1.5 cm. long, stigma large, diameter 5-6 mm.; flowering season February to April.

*External fruit characters.* Shape obtale, size 9.5-12.8 cm. ht., 12.5-17.2 cm. diam., wt. usually 2½-4 lb.; D/H index 1.3-1.47 (av. 1.39); colour light yellow; surface dull, papillat; base truncate or slightly depressed, slightly rugose; calyx more or less even, irregularly divided, medium coriaceous, 4-5 lobed; stem large, 2-2.8 cm. long; apex slightly depressed, occasionally even; navel occasionally present, navel about 10-12 per cent, somewhat large, both even and protruding, diameter of navel mark 2-2.4 cm.; areole absent, surface smooth, stylar scar even, style not persistent.

*Internal fruit characters.* Rind thickness 1.4-1.6 cm., firmness firm and brittle, adherence medium; oil glands number 10-18 per 1/4 sq. cm., very conspicuous, surface convex, size 1.5-2 mm. ht., 1.5-2 mm. diam., almost spherical or slightly oblate, aroma distinctly of pummelo, glandular layer about 1/6 th rind; mesocarp (albedo) light pink, spongy, bitter with a soupcon of sweetness; axis hollow or semi-hollow, mid-diameter 3.5-4 cm., segments 16-21, adherence medium to slight strong; septa coriaceous, tough, dorsal contour truncate, carpel size 7-8.5 cm. long, 4.5-5 cm. broad, 1.5-2.5 cm. thick; pulp-vesicles colour pink to deep pink, texture coarse, shape irregular, usually broad, fusiform, tips obtuse, pulp not easily separable, arranged in herringbone fashion, arising usually from dorsal side, flavour good; juice plenty, pink colour, wt. 13-16 oz. taste acidulously sweet, slightly bitter; seeds many, number 50-100 or more per fruit, size 1.5-2.3 cm. long, 1-1.1 cm. broad, 5-6 mm. thick, shape usually cuneate, plano-convex, with anastomosing lines on edges ridges or rough at the rounded end, testa cream white, tegmen light brown, chalazal spot brown, cotyledon white.

### *Kamrup (Rose coloured flesh)*

*Plant characters.* A tall plant, ht. 25-30 ft., densely foliaged, spreading branches, young branches pubescent; leaf large with prominent wing, deep green; mid-rib pubescent; fruiting season January to early February, one crop in a year.

*Flower characters.* Flower large, bisexual only; pedicel 1.5-2.5 cm. long. peduncle 3-5.5 cm. long, pubescent; calyx coriaceous, somewhat cupular, apex deltoid; corolla imbricate, 4-5 petals, shape oblong, apex obtuse, pure white, size 2.1-2.4 cm. long, 8. mm. 1.1 cm. broad; stamens 34-41, filaments some jointed and some free, white, glabrous, 1.2-1.4 cm. long, ovary inverted top shaped, size usually 5-7 mm. long, 4-5 mm. broad, style cylindrical, white, 1-1.1 cm. long, stigma large, diameter 3-4 mm.; flowering season February to April.

Kamrup (*C. grandis*)

*External fruit characters.* Shape turbinate to ovate-pyriform; large size 14.2-16 cm. ht., 13-1-15 cm. diam; wt. 1½-3 lb.; D/H index 0.90-0.94 (av. - 0.92); colour light yellow; surface finely papillate; base low collared, slightly depressed; stem 2.8-3 cm. long; calyx large, coriaceous, 3-4 lobed, tip acute; apex depressed, areole absent, stylar scar sunken or depressed in a small apical pit, style not persistent.

*Internal fruit characters.* Rind thick, thickness 1.7-2 cm., firmess firm and brittle, adherence medium strong; oil glands conspicuous, number 12-14 per 1/4 sq. cm., shape spherical to ellipsoid, size 1-2.5 mm. ht., 1-2 mm. dia., surface convex, aroma distinctly of pummelo, glandular layer 1/5 of rind; mesocarp (albedo) white, texture soft and spongy, taste slightly sweet with soupcon of bitterness; axis semi-hollow or hollow, diameter at the centre 1.5-2 cm. irregular in outline; segments 10-13, adherence medium to strong, carpel size usually 9.5-10.5 cm. long, 4-5 cm. broad, 1.7-2.4 cm. thick, septa thick, tough, closed or broken at the core, colour rose or with a dash of pink, dorsal contour slightly convex to truncate; pulp-vesicles rose coloured, texture coarse, separable without breaking the sac wall, somewhat loose, arranged in herringbone fashion, arising only from dorsal side of the carpel, vesicles irregular in shape and size, shape big ones usually rod-like or fusiform, tips obtuse, occasionaly slightly acute, flavour good; juicy, 6-8 oz., taste acidulously sweet slightly bitter; seeds many, 40-50 or more per fruit, cuneate, micropylar end flat, one side smooth and other side with anastomosing lines, rough at rounded end, 1.6-1.8 cm. long, 0.9-1 cm. broad, 5-6 mm. thick, testa white or slight cream white, tegmen light brown, chalazal spot sepia brown to brown cotyledon white.

***Khasi (Beef coloured pulp)***

*Plant characters.* A tall plant, ht. 25-30 ft., well branched, spreading; densely foliaged; almost thornless; young branches angular, pubescent; fruiting season October to mid December, one crop annually.

*Leaf characters.* Leaf large, broadly winged, coriaceous, deep green, shape ovate-oblong; size 10.5-12.2 cm. long, 5.8-7.8 cm. broad, apex attenuated, obtuse, margin annually crenate; petiole winged, obcordate, size 2.8-4.5 cm. long, 1.2-2.8 cm. broad, apex truncate to retuse, margin entire to slightly crenate; mid-rib pubescent on the lower side, veins slightly prominent on both sides; articulated.

*Flower characters.* Flower large, bisexual, glomerate (6-10), raceme; pedicel 1.5-2.2 cm. long, peduncle 1.2-5 cm. long, pubescent; calyx coriaceous, cup-shaped, usually 4-lobed; corolla imbricate, petals usually 4, white, shape oblong, size 2.2-2.5 cm. long, 0.9-1.3 cm. broad; stamens 30-38, anther 3-4 cm. long, filament 1.2 cm. long, polyadelphous, glabrous; ovary globose, size 4 mm. ht., 4-5 mm. diam. style cylindrical, white, 8-9 mm. long, including stigma, stigma large, dia. 4-5 mm., flowering season February to April.

*External fruit characters.* Shape almost spherical or globose; size large, variable normal size 13.5-14.5 cm. ht., 14.2-15.5 cm. diam., wt. 2½-4 lb.; D/H index 1.03-1.08 (av. 1.04); colour light yellow with a dash of pink; surface capillate; base evenly rounded, occasionally ribbed, calyx large, coriaceous 3-4 lobed, tips acute to somewhat obtuse; apex moderately or slightly depressed, smooth; areole absent, stylar scar even in a depressed cavity, style not persistent.

*Internal fruit characters.* Rind thick, thickness 2-2.4 cm. in median corss section, firmness firm and brittle, adherence medium strong; oil glands conspicuous, number 8-12 per ¼ sq. cm., primary glands somewhat spherical or slightly ellipsoid, 1-1.5 mm. long, 1 mm. broad, surface convex, glandular layer about 1/5 th of rind; mesocarp (albedo) red or deep pink, texture soft and spongy, taste, bitter with modicum of sweetness; axis usually hollow, irregular in outline in median cross section, diam. 1.8-2.5 cm; segments 12-16, adherence medium to strong, septa coriaceous, deep pink, usually not broken at axis, dorsal contour of carpels truncate to retuse; pulp-vesicles venison or beef coloured, texture coarse, easily separable, loose, arranged in herriangbone fashion, arising from dorsal contour, rarely from radial walls, irregular in size and shape, tips obtuse of slightly acute, not melting, shape fusiform and rod-shaped, size big ones 2-2.2 cm. long, 3-4 mm. broad, flavour good; juice medium, 4-6 oz. per fruit, colour pinkish, taste acidulously sweet, bitter principle not very marked; seeds usually cuneate, many, 80-140 per fruit, one side plano-convex, other side with ridges and furrows, micropylar end flat, very rough on rounded end, size 2-2.2 cm. long. 1.1-1.3 cm. broad, 5-6 mm. thick, testa cream white, tegmen light brown, chalazal spot Indian red, cotyledon white.

***Burni (White pulp)***

*Plant characters.* A tall to medium-sized plant, ht. 20-30 ft., erect; well branched; moderately dense foliage; almost thornless; young shoots angular and pubescent; season November to December, one crop in a year.

*Leaf characters.* Leaf large, coriaceous, glossy, deep green, blade ovate-oblong, margin almost entire, apex obtuse, size 10-14 cm. long, 6-8.5 cm. broad; petiole broadly or narrowly winged, size 2-4 cm. long, 0.5-2 cm. broad; shape obcordate or spathulate, apex truncate to retuse, margin very slightly crenate or entire; veins prominent on both sides mid-rib and veins pubescent on the lower side, articulated.

*Flower characters.* Flowers large, bisexual, clustered (3-10), raceme; pedicel 1.5-2 cm. long, peduncle 1.2-4.5 cm. long, pubescent, calyx coriaceous, cupular, usually 4-lobed; corolla imbricate, petals 4-5, oblong, apex obtuse, size 2.1-2.5 cm. long, 0.8-1 cm. broad; stamens 25-36, anther 3-4 mm. long, filaments polyadelphous, glabrous, white, 0.9-1.2 cm. long; ovary almost, spherical, size usually 4 mm. ht., 4 mm. diam., style white, sylindrical, 8-9 mm. long including stigma, stigma capitate, diameter 4-5 mm.; flowering season March to April.

*External fruit characters.* Shape spherical or globose, size medium, variable, normal size 11-13.2 cm. ht., 11-13.5 cm. diam.; wt. 1½-3 lb.; D/H index 0.98-1.05 (av. 1.01); colour light yellow; surface smooth, pubescent on tough, finely papillate; base evenly rounded, rarely rugose; stem 1.5 cm. long; calyx medium coriaceous, irregularly divided, usually 3-4 lobed, tips either acute or obtuse; apex slightly depressed or almost rounded; areole absent, stylar scar depressed in a small apical pit, style not persistent.

**Zemabawk (*C. grandis*)**

*Internal fruit characters.* Rind thick, thickness 1-2 cm. in median corss section, firmness firm and brittle, adherence somewhat strong; oil glands conspicuous, number 12-14 per ¼ sq. cm., shape somewhat spherical or oval, 1.5-2 mm. long, 1-2 mm. broad, surface convex, aroma distinctly of pummelo glandular layer 1/5th of the rind; mesocarp (albedo) white, soft and spongy, taste mildly sweet with soupcon of bitterness; axis usually semi-hollow, mid diameter 1.5-2 cm.; segment 12-14, adherence medium to strong, septa coriaceous, tough and leathery, white, both broken or closed at axis, dorsal contour convex to truncate; pulp-vesicles white, texture coarse, easily separable, loose, arranged in herringbone fashion, arising from dorsal side of the carpels, shape and size irregular, usually fusiform, tips obtuse, size big ones 1.5-2.2 cm. long, 2-3 mm. broad, flavour good; juice medium, wt. 5-8 oz. per fruit, taste almost sweet no bitterness; seeds 10-30 per fruit, irregular in shape, usually cuneate with broad and flat micropylar end, surface somewhat rough, plano-convex, size 1.5-1.7 cm. long, 1-1.1 cm. broad, 5-6 mm. thick, testa white, tegmen very light brown, chalazal spot brown, cotyledon white.

This is a desirable pummelo, worth propagating as a horticultural variety. It has been found to be one of the most delicious varieties of pummelo described in this work.

### *Zemabawk (Honey yellow pulp)*

*Plant characters.* A medium-sized plant, ht. 18-20 ft., moderately spreading branches; densely foliaged; almost thornless; young shoots, imperceptibly pubescent; fruiting season December to January, one crop in a year.

*Leaf characters.* Leaf medium to large with narrow petiolar wing size 7.5-13.6 cm. long, 3.7-6.5 cm. broad with petiolar wing 1.4-2.4 cm. long, 2-7 mm. broad; shape (blade) lanceolate to broadly lanceolate, coriaceous, dark green, articulated, margin crenate, apex slightly acute, occasionally bifid; petiole spathulate to rundimentary, apex obtuse to truncate, margin more or less entire.

*Flower characters.* Flower bisexual, usually glomerate (3-8); pedicel; calyx, medium coriaceous, stellate, deltoid, apex acute, ciliate; corolla imbricate, petals 4-5, pure white, oblong, size 1.7-2.1 cm. long; 7-9 mm. broad; stamens 20-24, filaments polyadelphous, white, glabrous, 0.9-1.4 cm. long, anther 2-3 mm. long; ovary 3-4 mm. long, 3 mm. broad, style cylindrical, glabrous, white, 8-9 mm. long including stigma, stigma capitate, 3 mm. diam : flowering season March to April.

*External fruit characters.* Shape usually turbinate, occasionally ovate pyriform, size 11-13.8 cm. ht., 10.8-12.5 cm. diam.; wt. 1-2 lb.; D/H index 0.84-1.01 (av. 0.93); colour light yellow; surface almost smooth the rugose, minutely papillate, pitted; base slightly depressed, very low-necked or even; calyx depressed, irregularly divided, tips acute or obtuse; stem 0.8-1.2 cm. long; apex slightly depressed to almost rounded, areole indistinct, stylar scar present, even in a small apical pit, style not persistent.

*Internal fruit characters.* Rind somewhat thick, thickness 1-1.3 cm. in median cross section, firmness firm, adherence medium strong; oil glands conspicuous, number 14-18 per ¼ sq. cm., shape variable, usually spherical, elliptical, occasionally pyriform, size 1.5-3 mm. long, 1-2 mm. broad, glandular layer 1/3 rind; mesocarp (albedo) white, thick, soft and spongy, taste slightly sweet with soupcon of bitterness; axis semi-hollow, pithy, mid-diam. 1.2-2 cm., septa closed at axis; segments 9-10 (usually 10), adherence medium to slightly strong; septa coriaceous, white, tough and leathery, carpel size 7-9.5 cm. long, 3.4-4 cm. broad, 1.8-3 cm. thick, dorsal contour truncate to retuse; pulp-vesicles somewhat light yellow to honey yellow, texture coarse, very soft and melting, arising from dorsal and lower margin of carpellary walls, shape irregular, usually fusiform, big ones 1.2-2 cm. long, 2-5 mm. broad, stalked, 1-1.5 cm. long, tip obtuse, closely packed, flavour good; juice abundant, 7-10 oz. (usually 8 oz. per fruit), colour light yellow or somewhat honey yellow, taste acidulously sweet, bitter principle not marked; seeds many, 20-42 per fruit, size large, usually 1.3-1.5 cm. long, 8-9 mm. broad, 5-6 mm. thick, more or less cuneate, broad and flat at the micropylar end, one side smooth, other side with anastomosing lines, rounded end distinctly ridged, testa cream white, tegmen light brown, chalazal cap sepia brown, cotyledon white.

The variety *Zemabawk* is almost akin to grapefruit (*C. paradisi* Macf.) in many diagnostic details and can probably be treated as a horticultural variety under the grapefruit. The authors have, however, dealt in under the pummelo as this variety has no commercial importance as yet to justify a varietal rank under the grapefruit. It is not well distributed and has probably originated as a chance seedling from the pummelo. It is being propagated by vegetative means.

### *Aijal (Light yellowish white pulp)*

*Plant characters.* Plant medium to tall, ht. 18-20 ft.; moderately spreading branches; densely foliaged; almost thornless; young growing shoots obscurely pubescent; fruiting season January to March, one crop in a year, medium fruiter.

**Aijal (*C. grandis*)**

*Leaf characters.* Leaf medium to large with rundimentary to narrow petiolar wing, size 7-17 cm. long, 5.5-10.5 cm. broad with petiole measuring 1.2-2 cm. long, 0.4-1.2 cm. broad, blade ovate-oblong, coriaceous, dark green, articulated, margin crenate, apex attenuated, obtuse, emarginate; petiole winged, usually spathulate, rarely obcordate, margin entire, apex obtuse or slightly truncate; veins strong on upper side of the leaf.

*Flower characters.* Flowers bisexual, either solitary or in clusters (1-6), cymes; pedicel 5-8 mm. long, peduncle 2-5 cm. long, minutely pubescent; calyx medium thick, more or less stellate, deltoid, apex acute, ciliate, greenish white; corolla imbricate, petal 4-5, pure white, oblong, apex either slightly acute or almost obtuse, size 1.8-2.3 cm. long, 0.8-1.2 cm. broad; stamens 23-27, filament white, polyadelphous, occasionally 1 or 2 free, glabrous, 1-1.5 cm. long, anther 2-3 mm. long; ovary almost spherical to oblate, size 3-4 mm. long, 3-4 mm. broad, style cylindrical, white, glabrous, 7-8 mm. long, stigma capitate, 3 mm. diam.; flowering season March to April.

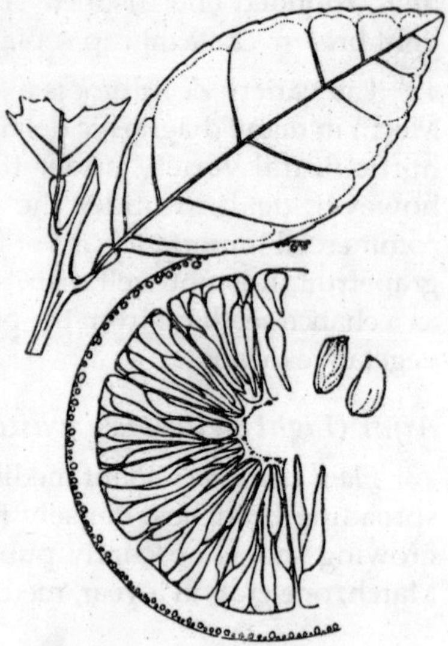

**Zemabawk *(C. grandis)* showing outline of leaf and internal structural of fruit**

*External fruit characters.* Shape round-oblate, size medium, 6.6-10 cm. ht., 10-11.1 cm. diam.; wt., 12-20 oz. (usually 15 oz.); D/H index 1.20-1.53, (av. 1.25); colour bright yellow; surface almost smooth, pitted and also finely papillate; base almost rounded or even, very minutely ribbed; calyx slightly depressed, irregularly divided, tip obtuse; stem short to medium, 1-1.2 cm. long; apex obtuse, areole absent, stylar scar present, even in a very small depression, style not persistent.

*Internal fruit characters.* Rind medium thick, thickness 6-8 mm., firmness medium, adherence medium strong; oil glands conspicuous, number 15-20 per 1/4 sq. cm., shape usually oval or elliptical, occasionally pyriform, size 2-3 mm. long. 1-2 mm. broad, oil medium, aroma weak, glandular layer almost 1/2 rind; mesocarp (albedo) white, bitter with soupcon of sweetness; axis semi-hollow or pithy, diam, at the centre 1.5-2.1 cm., carpel closed at axis; segments 10-13, (usually 12-13), adherence strong, septa white, chartaceous, tough and leathery, carpel size 6.07 cm. long, 3-3.5 cm. broad, 1.5-2.5 cm. thick; pulp-vesciles colour somewhat light yellowish white, texture coarse, very soft and melting, vesicles arising from dorsal and lower margin of carpellary walls, shape irregular, usually fusiform, tips slightly acute to obtuse, stalked; juice abundant, 6-8 oz. per fruit, colour very light yellow, taste acidulously sweet, bitterness not marked; seeds many, 15-30 per fruit, size medium large, 1.5-1.6 cm. long, 8-9 mm. broad, 5-6 mm. thick, shape almost cuneate, broad and flat at the micropylar end, one side smooth, anastomosing lines on other side, rounded and slightly rough or ridged, testa white, tegmen light brown, chalazal spot brown, cotyledon ivory white.

The variety *aijal* compares well with the standard grapefruit. It is of rare occurrence and has as yet attened no commercial status. This might also have originated as a chance seedling. The fruit is not so sweet under Burnihat conditions, though the look is attractive. In suitable soil and climatic conditions especially in the southern slope of the Khasi Hills, the variety may improve in quality. It is being propagated now by vegetative means at the Citrus Fruit Research Station, Burnihat to establish a clonal variety. It has provisionally been placed under the pummelo (*C. grandis*).

## The Sour Pummelos (*Citrus megaloxycarpa* Lushington)

*Origin and name.* The sour pummelo, commonly known as *amilbed* in Western India is an intensely acid fruit of very big size, resemblig outwardly the proper pummelo (*C. grandis*, Osbeck) All evidences tend to show that it is indigenous India as similar fruits do not seem to have been reported to exist elsewhere. The earliest reference to this fruit was made in Babar's Memoir. There are several forms of this, met with in Assam, popularly called in Assamese as *bor-tenga* which in its English rendering means a giant form of sour citrus.

*Nomenclature.* The position of the *amilbed* or *bor-tenga* is very anomalous. Because of the apparent resemblance of these fruit with the proper pummelo it is sometimes confused though it is singularly distinctive from the latter by its tree habit, atrophied petiolar wing, light purple coloured flowers, intensely sour juice and also by the purple coloured chalazal spot. That this is more antique than the proper pummelo in India is evident from the fact that the latter was not mentioned in Babar's Memoir though as a quality fruit it is unquestionably superior to *amilbed*. Bonavia (1890) states 'It is, therefore, resonable to infer that in Babar's time, or about 300 years ago the pummelo proper, as we see it to-day, was not known in India.' Bonavia though seemingly impressed by the specific differences exhibited by these two classes of fruit considered *amilbed* only as a variety of the proper pummelo without advancing any taxonomical reasons for its amalgamation. Lushington (1910) was the first to assign a specific name *Citrus megaloxycarpa* to *amilbed* as distinct from the pummelo (*C. grandis*). Tanaka (1937 a) obviously fiding this basically different from the proper pummelo recognized *amilbed* by another specific name *C. pennivesiculata,* although Lushington (1910) coined this name for a variety within the species *C. megaloxycarpa.* It is really strange that in the long list of Assam *Citrus* examined and identified by Tanaka none of the different forms of the *amilbed* group (*bor-tenga*) so common in Assam, features even once.

It is really surprising that the *amilbed* which has been breeding true to type without any change of its inherence characters almost unaided through the passage of several centuries and which is well distributed in India, has not received the attention it deserves from the taxonomists. In view of the basic differences that exist between these two classes of fruits it seems to be reasonable to give the *amilbed* or the Sour Pummelo of India a specific rank and the name *C. megaloxycarpa* Lush. being the earliest in the field should be recognized.

**Key to Varieties of the Sour Pummelo.**

A. Plant medium tall, (10-20 ft.), spreading; leaves usually oblong to ovate-oblong, almost wingless, or narrowly winged; young shoots either glabrous or pubescent; flower usually slightly tinged; fruit large (10-24 cm. diam.); juice abundant, intensely acid; seeds usually cuneate, chalazal cap Indian red or purple.

B. Fruit spherical to turbinate; surface slightly warty or undulating; pulp white.

$V_1$..............*Bor-tenga*

BB. Fruit spherical or globose; surface smooth, pubescent on touch.

$V_2$..............*Hukma-tenga.*

BBB. Fruit oval to slightly spherical; surface unusually very warty and undulating.

$V_3$..............*Holong-tenga.*

BBBB. Fruit oblong with broad mamilla at apex; surface smooth and finely papillate; pulp greenish white.

$V_4$..............*Jamir-tenga.*

## *Bor-tenga*

*Plant characters.* A medium-sized plant, ht. 10-12 ft., straggling branches, moderately dense foliage; very thorny; young parts glabrous, fruiting season October to middle of January, one crop in a year.

**Bor-tenga (*C. Megaloxycarpa*)**

*Leaf characters.* Leaf large with rudimentary or narrow wing, blade deep green, shape ovate-oblong, margin crenate, apex obtuse, emarginate, size 9.5-16.3 cm. long, 6-10.2 cm. broad; petiole 1.5-2.1 cm. long, 0.5-1 cm. broad, shape spathulate, articulated.

*Flower characters.* Staminate and pistillate, proportion of pistillate flowers more than the staminate ones, cymes; pedicel 0.6-1.1 cm. long, peduncle 1-3 cm. long, glabrous; calyx coriaceous, sepals tinged; corolla imbricate, petals 4-5, oblong, purple, size 2.2-3 cm. long 0.8-1.3 cm. broad; stamens 30-38, filament polyadelphous, glabrous, white, 1.1-1.9 cm. long; ovary oblong, size 5-6 mm. long, 4 mm. broad, style cylindrical, white, turns into light purple, 1.1-1.3 cm. long, stigma capitate, 4-5 mm. diam.; flowering season March to April.

*External fruit characters.* Shape almost spherical to slightly turbinate; size variable, normal ones 15-20.1 cm. ht., 15-18.2 cm. diam.; wt. 2-5 lb.; D/H index 0.90 to 1.01 (av. 0.97); colour light yellow surface slightly smooth to sub-warty, rather uneven, prominently foveolated; base broadly necked, ribbed, depressed; calyx sunken inside collar, irregularly divided, 4-5 lobed; apex moderately depressed, occasionally somewhat truncate; areole indistinct, areolar area slightly rough, very slightly rugose or wrinkled, stylar scar present, diam 1-1.5 mm.; depressed in a small apical pit, style persistent.

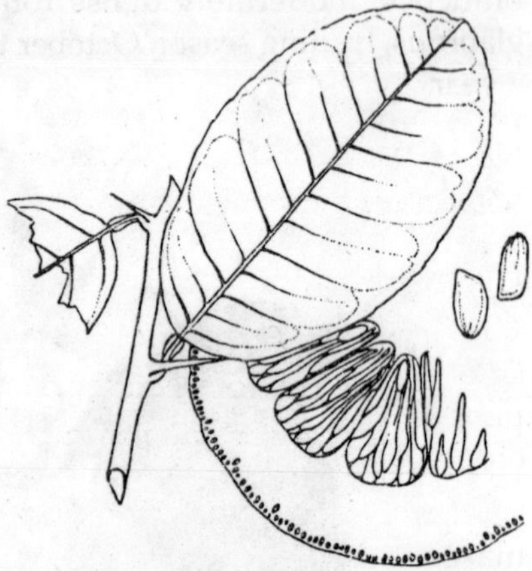

**Bor-tenga (C. megaloxycarpa) showing outline of leaf and internal structure of fruit**

*Internal fruit characters.* Rind thick, thickness 1.5-3.5 cm. in median cross section, firmness firm and brittle, adherence medium; oil glands large, size. 2-3 mm. long, 1-2 mm. broad, number 8-12 per 1/4 sq. cm., shape oval. aroma distinct, glandular layer 1/5 rind; mesocarp (albedo) chalky white spongy, soft, taste very sweet with soupcon of bitterness; axis usually hollow diam. at the centre 1.4-3.5 cm.; segments 10-14 (usually 13), adherence medium; septa medium coriaceous, tough, white, sometimes broken at axis, dorsal contour usually convex to truncate; pulp-vesicles white, texture coarse, somewhat melting, arranged in herringbone fashion, arising from dorsal contour only, shape fusiform, size big, 1.7-2.5 cm. long, 2-4 mm. broad, stalked; juice abundant, 10-16 oz. per fruit, colour white, taste very sour, not bitter; seeds many, 50-70 per fruit, size 1.5-1.8 cm. long, 0.9-1.1 cm. broad, 6-7 mm. thick, shape ovate to cuneate, micropylar end with fin-like projection of testa, surface almost smooth on both sides, slightly wrinkled on the round end, testa cream white, tegmen light brown, chalazal cap purple, cityledon white.

This variety is usually met with in Assam Valley plains. The fruits are very sour and worthless. It must have long existed in Assam as it is very common in villages as a curiosity plant. It has been referred to by Robinson (1841) as *bor-muri-tenga.* Anderson of Sibsagar included *bor-tenga* (*bar-tenga*) in the list of Citrus fruits as reported and reproduced by Bonavia (1890) in pp. 334, and in Plate No. CCXXXVII. The *bor-tenga* is known as *pani-matu* in Cachar and *soh-shy* in the Khasi Hills.

### *Hukma-tenga*

*Plant characters.* A medium-sized plant, ht. 15-20 ft., erect; spreading; moderately dense foliage; almost thornless; young shoot pubescent, angular; fruiting season October to midle fo January, one crop in a year.

*Leaf characters.* Leaf large, shape ovate-oblong, dark green, margin crenate, apex obtuse, size 12.5-15.5 cm. long, 7-9.5 cm. broad; petiole wingless, 1.2-1.5 cm. long, 4 mm. broad; mid-rib pubescent on lower side, articulated.

*Flower characters.* Bisexual only, raceme, clustered, (2-8); pedicel 0.8-1.3 cm. long, peduncle 0.6-5 cm. long, pubescent; calyx medium thick, deltoid, apex pointed, ciliate, 4-5 lobed; corolla imbricate, petals 4-5, colour very light purple outside, ephemeral, oblong, size 2.2-2.5 cm. long, 8 mm. -1 cm. broad; stamens 19-24, filaments usually free, occasionally jointed by two together, white, glabrous, 1.1-1.5 cm. long, anther 4-6 mm. long; ovary globose, pubescent, size 4-5 mm. long, 4 mm. broad, style cylindrical, pubescent, 1.1-1.4 cm. long includiing stigma, stigma capitate, 4-5 mm. diam.; flowering season March to April.

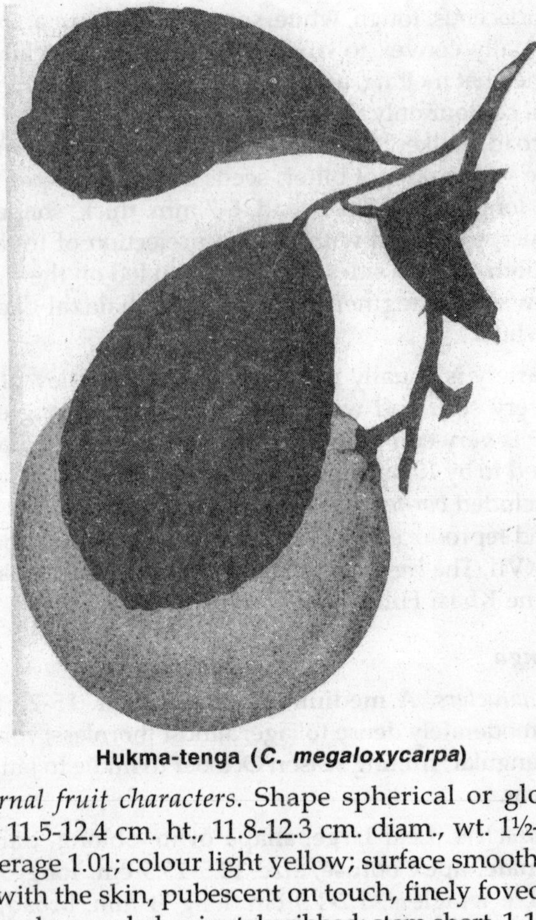

**Hukma-tenga (*C. megaloxycarpa*)**

*External fruit characters.* Shape spherical or globose; size variable, 11.5-12.4 cm. ht., 11.8-12.3 cm. diam., wt. 1½-2 lb.; D/H index average 1.01; colour light yellow; surface smooth, oil glands in flush with the skin, pubescent on touch, finely foveolated; base almost evenly rounded, minutely ribbed; stem short, 1-1.5 cm. long, calyx medium coriaceous, persistent, 4-5 lobed; apex evenly rounded, areole indistinct, stylar scar slightly depressed, style not persistent.

*Internal fruit characters.* Rind thick, thickness 0.8-1.1 cm. in median cross section, firmnes firm and brittle, adherence strong; oil glands number 9-12 per 1/4 sq. cm., primary glands somewhat spherical or oval, occsionally pyriform, size 2-2.5 mm. long, 2 mm. broad, glandular layer 1/5 rind; mesocarp (albedo) white, spongy, taste very slightly sweet with soupcon of bitterness; axis solid or hollow, diam. at the centre 1-1.2 cm.; segments 9-12, adherence strong; septa medium coriaceous, tough, white, broken or closed at axis, dorsal contour convex to truncate; pulp-vesicles white, texture somewhat coarse, melting, arising from dorsal contour and lower wall of septa, fusiform, big ones size 0.7-1.1 cm. long, 4 mm. broad, flavour not good; juice abundant, 9-10 oz. per fruit, colour white, very sour, free from bitter principle; seeds large, many, 25-60 per fruit, more or less cuneate, anastomosing lines on both sides, rough on rounded end, micropylar end flat, testa cream white, tegmen light brown, chalazal spot purple, cotyledon white.

The variety is not commonly grown and has no commercial importance. The specimen has been collected from Burnihat where it is grown as curiosity plant.

### *Holong-tenga*

**Holong-tenga (*C. megaloxycarpa*). Mark peculiar roughness of the rind.**

*Plant characters.* A medium-sized tree, ht. 14-18 ft., spreading; thorny; densely foliaged; fruiting season November to February, one crop in a year.

*Leaf characters.* Leaf coriaceous large with very small wing or almost wingless; (blade) ovate-oblong to broadly lanceolate, deep green, margin crenats to slightly serrate, apex acute, emarginate, size 12-21 cm. long, 6-12 cm. broad; petiole 1-2.5 cm. long, 0.4-1.7 cm. broad, articulated; mid-rib pubescent.

*Flower characters.* Bisexual, rarely staminate, clustered (2-6), large, raceme; pedicel 1.2-1.5 cm. long peduncle 0.5-3 cm. long, pubescent; calyx cupular, greenish white; corolla imbricate, petals 4 oblong, pure white, size 3-3.5 cm. long, 1.2-1.5 cm. broad; stamens 27-37, filament white, glabrous, polyadelphous, occasionally 1 or 2 free, 1.6-2 cm. long, anther 5-6 mm. long; pistil more or less club-shaped; ovary oblong, size 7-9 mm. long, 5-6 broad, style cylindrical, white, 1.3-1.5 cm. long, stigma broad, 5-6 mm. diam.; flowering season February to March.

*External fruit characters.* Shape oval to slightly spherical, 16-28 cm. ht., 15-24 cm. diam.; weight 3-8 lb.; D/H index 0.79-0.93, (av. 0.85); colour lemon yellow; surface unusually warty and undulating; oil glands very much conspicuous, distinctly foveolated; base deeply depressed, ribbed, furrows extending through the collar; calyx sunken in the cavity, irregularly divided, sepal 8-9 mm. long, 0.7-1.2 cm. broad, coriaceous, reflexed upwards, tips rounded; stem 1.2-2.5 cm. long; apex rounded to flat and depressed; areole indistinct, areolar area very rough and indistinctly ribbed, stylar scar present, diam. 3-5 mm. depressed in a apical pit, style not persistent.

**Holong-tenga (*C. megaloxycarpa*)**

*Internal fruit characters.* Rind very thick, thickness 2.5-2 cm. firmness firm and brittle, adherence medium; oil glands very large, number 5-7 per 1/4 sq. cm., shape usually ellipsoid, occasionally pyriform, size 2.5-6 mm. long 1-3 mm. broad, aroma distinct, glandular layer slightly more than 1/5 rind; mesocarp (albedo) chalky white, spongy, soft, taste very slightly sweet with soupcon of bitterness; axis hollow, diameter at the centre 1.5-3 cm., segments 11-14 (usually 13), adherence medium, septa coriaceous, tough and leathery, broken at axis, dorsal contour convex, truncate, or concave; pulp-vesicles white, texture coarse, somewhat melting, arranged in herringbone fashion, arising from dorsal and lower portion of the carpellary walls, shape usually fusiform, size big ones 1.5-2.5 cm. long, 3-6 mm.broad, stalked, 1-1.6 cm. long,tip obtuse, juice abundant, 12.7 oz. per fruit, colour white, taste very sour, not bitter; seeds few, sometimes seedless, 0-12 seeds per fruit, size 1.6-1.7 cm. long, 0.8-1 cm. broad, 5-6 mm. thick, somewhat cuneate, one side smooth, other side ridged, rough towards rounded end, micropylar end flat, testa cream white, tegmen brown or light brown, chalazal cap deep purple, cotyledon white.

This is a very common citrus fruit in Assam Valley plains. The fruit is very sour and valued for its alleged medicinal properties. The popular belief is that if a needle be thrust inside the fruit and kept for long it gets dissolved. But the chemical analysis of the fruit juice shows that it is not so sour as it is believed to be. This fruit has been referred to by Robinson (1841) as *bor-holong-tenga.*

There is an acidless form of this variety having no other morphological differences, recently detected in Nowgong (Assam) and is known as *mitha-holong-tenga.* As this is just an isolated plant, it cannot be recognized as a variety but can only be considered as a chance hybrid or a sport.

### *Jamir-tenga*

*Plant characters.* Plant small, ht. 10-12 ft., spreading; moderately dense foliage; young branches angular; pubescent; very thorny, slender, sharp; plant not very hardy; fruiting season December to February, one crop annually.

*Leaf characters.* Leaf medium to large with moderately broad wing, occasionally rudimentary or narrowly winged; blade broadly lanceolate, apex acute, emarginate, margin more serrate than crenate, medium coriaceous, deep green; petiole obcordate to spathulate, margin minutely crenate, apex truncate to slightly retuse; size (blade) 8-17 cm. long, 3.5-8.0 cm. broad with petiolar wing 1.5-2.5 cm. long, 4 mm.-1.6 cm. broad; veins somewhat prominent on both sides, mid-rib pubescent on lower side.

*Flower characters.* Flower bisexual only, clustered (2-8), cymes; pedicel 0.8-1.6 cm. long, peduncle 0.8-1.8 cm. long, pubescent; calyx coriaceous, greenish whtie, cupular, corolla imbricate, petals

4-5, colour in bud stage slightly purple tinged, ephemeral, white in open petals, oblong, size usually 2.2-3 cm. long, 0.7-1 cm. broad; stamens 28-38, filament polyadelphous, also some free, glabrous, white, 1.2-1.5 cm. long, anther 5-6 cm. long; ovary ovate-oblong, size 5 mm. long, 3-4 mm. broad, style white, turns, into light purple, cylindrical, 1-1.2 cm. long, stigma capitate, diam. 3-4 mm.; flowering season March to April.

**Jamir-tenga (*C. megaloxycarpa*)**

*External fruit characters.* Shape oblong or somewhat lemon shaped, size 11-20.8 cm. ht., 9-16 cm. diam; weight 1½-5 lb.; D/H index 0.72-0.83 (av. 0.78); colour lemon yellow; surface smooth, very finely papillate, dull; base low-collard, slightly depressed; calyx large, irregularly divided, medium, coriaceous, tips obtuse; stem 1.5-2.5 cm. long; apex usually mamillate, occasionally almost rounded; areole indistinct, areolar area radially ribbed, style partially persistent, occasionally depressed in a small apical pit.

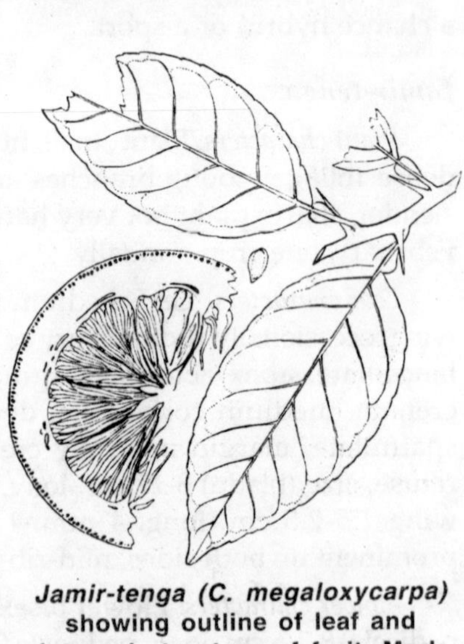

***Jamir-tenga (C. megaloxycarpa)* showing outline of leaf and internal structure of fruit. Mark the needle shaped pulp-vesicles**

*Internal fruit characters.* Rind thick, thickness 0.7-2 cm. (usually 1-1.5 cm.), firmness firm and brittle, adherence medium to strong; oil glands conspicuous, number 10-18 per 1/4 sq. cm., shape usually ellipsoid, spherical, or slightly oblate, size 0.5-3 mm. long,

0.52 mm. broad, closely compressed, oil medium, aroma weak, distinct, glandular layer 1/8th rind; mesocarp (albedo) white, soft, spongy, taste bitter with soupcon of sweetness; axis solid, diam. at base 6-8 mm., at the centre 0.8-1.2 cm., apical end 0.8-1.2 cm.; segments 9-11 (usually 10), adherence strong; septa medium coriaceous, tough, dorsal contour truncate or retuse; pulp vesicles greenish white, texture medium coarse to slightly fine, slender, melting, arising from dorsal and lower half of the carpellary walls,size big ones 1.2-3.5 cm. long. 2-5 mm. broad, shape fusiform, tip acute, stalked, 1-1.6 cm. long; juice abundant, 10-18 oz. per fruit, colour slight greenish white, very sour, free from bitterness; seeds 6-28 per fruit, size 1.2-1.6 cm. long, 0.8-1 cm. broad, 4-7 mm. thick, shape somewhat obovoid, occasionally cuneate, usually minutely ridged, one side smooth, rounded end rough, micropylar end flat, testa cream white, tegmen brown or light brown, chalazal spot purple, cotyledon white.

This is a common plant of the Assam Valley plains and has been seen growing in the hills. The fruits are highly acidic and are esteemed for medicinal properties; otherwise these have no value. In the *Account of Assam* by Robinson (1841) this variety has been mentioned under his so called acid lime group. The fruit bears a close resemblance to *hulinga-muri* of Assam, referred to by Bonavia in his Plate No. CCXXXVI. It is highly susceptible to gummosis and is not long-lived.

# Chapter 3
# SUBGENUS PAPEDA

## Section *Papedo citrus*

### The Ichang Papeda (*Citrus ichangensis* Swingle)

*Origin and name.* This remarkable species is considered to be a native of south-western China, occuring in Hupeh, Szechwan, Kweichow and Yunnan provinces. In Assam, it is found to grow absolutely in a wild state in the Barail range of the Naga Hills at an elevation of about 5000-6000 ft. The place from where the specimen has been collected is near a village called Khonoma about 35 miles from Kohima, the district headquarters of the Naga Hills.

This is known as *ketsa-shupfa* in the Angami language of the Naga Hills. The seedling progenies of this specimen have been recently planted at the Citrus Fruit Research Station, Burnihat but the growth of the plants appears to be very slow under Burnihat condition.

*Nomenclature.* Swigle (1913) thoroughly described *Ichang papeda* and raised it to the rank of a distinct species *Citrus ichangensts.* The botanical descriptions of the specimen collected in Assam conform essentially to those of one described by Swingle (1946). In his descriptions the characters like shape and size of the fruit except for diameter shape and size of vesicles, pitted or dented spots over the seeds towards the rounded end and colour of the chalazal spot are wanting, probably because the speciemens. All these aforesaid important characters, besides others, have been incorporated in this work. Taking all the botanical details into consideration it seems certain that the *ketsa-shupfu* of Assam indistinguishably represents the type species of the *Ichang papeda* and as such the authors accept the botanical name *Citrus ichangensis* for *ketsa-shupfu* of the Naga Hills.

## Key to the Ichang Papeda

A. A medium-sized tree, 15-20 ft. ht.; leaf (blade) ovate-acuminate or ovate-lanceolate; size 5.5-6 cm. long, 2-2.5 cm. broad; petiole broadly winged. obviate or oblong-spathulate, size 4-6 cm. long, 1.8-2.8 cm. broad; veins runing parallel to one another particularly in petiolar wing.

B. Fruit small, (1.4-5.8 cm. diam.) shape oblong; surface finely papillate, smooth; pulp-vesicles small, globose, contain (acrid) oil droplets, sour and acrid taste; seeds very large, thick, shape more or less triangular to almost rectangular, dented spots towards rounded end, chalazal cap brown covering an area about 1.7 cm. long, 1.4 cm. broad, cotyledon thick, white

$V_1$.........*Ketsa-shupfu*

## *Ketsa-shupfu*

**Ketsa-Shupfu (C. ichangensis)**

*Plant characters.* A medium-sized tree, 15-20 ft. ht., young growing branches angular, glabrous; spinous, sharp, 1.7-3 cm. long, occasionally more; leaf broadly winged, dark green, articulated. blade ovate-lanceolate or ovate-acuminate, size 5.5-6 cm. long, 2-2.5 cm. broad, apex somewhat acuminate or caudate, occasionally emarginate, margin minutely crenate; petiolar wing obovate or oblong-spathulate, size 4-6 cm. long, 1.8-2.8 cm. broad, apex evenly rounded, narrowed abruptly at the base, margin almost entire; veins slightly rominent on the underside running closely parallel to one another particularly in petiolar wing; (flower not seen).

**Ketsa-shupfu (*C.ichangensis*)**

*External fruit characters.* Shape oblong with gradual tapering at base and apex; size 6.8-7.2 cm. ht., 4-5.8 cm. diam.; DH index 0.81; colour light lemon yellow; surface smooth, finely papillate, also minutely foveolated; base with very low-collar, very minutely rugose; calyx thin, small, divided division indistinct; apex gradual tapering with a broad papilla; areole indistinct, areolar area very minutely rugose; stylar scar even, style not persistent.

***Ketsa-shupfu (C. ichangensis)*** **showing the outline of leaf and internal structure of fruit. Mark the big size of seeds and small and globular shape of pulp-vesicles.**

*Internal fruit characters.* Rind thin, thickness 3-4 mm., adherence medium strong; oil glands conspicuous, surface convex, glandular layer about 1/3 of the rind, number 18-20 per 1/4 sq. cm. aroma of the crushed glands resembles that of the green mango (*Mangifera indica*), distinct; mesocarp (albedo) white, tasteless, texture soft, spongy; axis solid or semi-hollow, diameter at the centre 0.81 cm.; segments 7.9, adherence of carpel medium strong; septa chartaceous, dorsal contour convex to slightly truncate, carpel practically filled up by the big seeds, 2-3 in each carpel; pulp-vesicles crystal white, loosely packed near the dorsal side of the carpels, separable, vesicles small, innumerable in each carpel, shape uniformly globose or somewhat ellipsoid, resembling the size and shape of sago, apex obtuse, size 3-5 mm. long, 2-4 mm. broad, with (filiform) thread like stalk, 2-4 mm. long, vesicles contain droplets of acrid oil, taste sour and acrid; juice scanty, not measurable; seeds 22-30 per fruit, very large, very thick, shape somewhat triangular to almost rectangular forming 3 to 4 planes, dented spots on the rounded end, micropylar end flat or occasionally broadly pointed, size 1.2-2 cm. long, 1-1.8 cm. broad, 1-1.2 cm. thick, testa white, tegmen cream white, chalazal spot light brown, covering an area cf 1-1.7 cm. long, 0.8-1.4 cm. broad, cotyledon very thick, white seed monoembryonic.

The *Ketsa-shupfu* is not a cultivated fruit and is absolutely inedible. It is reported to be very much cold-resistant. The fruits practically contain no juice and have no commercial importance. Its value as a root-stock has not yet been ascertained.

## The Khasi papea (*Citrus latipes* (Swingle) Tanaka)

*Origin and name.* The species *Citrus latipes* was founded on a specimen collected from the Khasi Hills of Assam. It may be considered as endemic to Assam. This species is also reported to occur in Northern Burma.

The species is represented by the *soh-kymphorshrieh* of Assam which in Khasi means the fruits for the monkey. As a matter of fact the fruits are so sour that even monkeys feel chary to eat. It is a semi-wild plant usually found (2000 to 6000 ft. altitudes) in places like Shillong, Mawprem, Moflong, (Mawphlang), Pynursla and in places down Cherrapunji. It is also reported to occur in the higher elevations of the Naga Hills where other wild citrus are often met with.

*Nomenclature.* Swingle (1913) classified this under the species *C. inchangensis*, sub-species *latipes* which subsequently was changed by him to a separate species, *C. latipes*, owing to its basic taxonomical difference with *C. ichangensis*. Tanaka (1937) identified *soh-kymphor* of the Khasi HIlls as *C. latipes*. In course of present investigation in Assam another form of the species has been recognized in *soh-shyrkhoit* which in some places of the Khasi Hills is also known by the name of the *soh-kymphorshrieh* referred to above. The *soh-shyrkhoit* is distinctive from the type species by its large sized fruits and by some slight differences in leaf and floral characters, described fully in the text. Both have been treated here as varieties under *Citrus latipes*.

## Key to Varieties of the Khashi Papeda (Citrus Latipes)

A. Plant medium to tall; densely foliaged, blade ovate-lanceolate; petiolar wing oblong-spathulate, either equal or slightly longer than the lamina.

B. Flowers usually glomerate, 4-merous only. slight purple tinged (ephemeral), stamens 20-30 filaments free.

C. Fruit almost globular; size 4.3-6.5 cm. ht., 4.8-6.6 cm. diam; juice scanty, very acid, pulp fusiform, somewhat greenish white; segments 9.11; seeds many, 30.60 per fruit, chalazal cap Indian red, cotyledon white.

$V_1$........*Soh-kymphorshrich.*

CC. Fruit shape oblate, depressed at base and apex, large; size 7.3-8.5 cm. ht., 8-9.8 cm. diam; segments 9.15; juice plenty (3-4 oz.), sour

$V_2$......*Soh-shyrkhoit.*

### *Soh-kymphorshrieh*

*Plant characters.* A medium to tall tree, 20-30 ft. ht., spreading densely foliaged; thorny, stout big ones 3-6 cm. long; fruiting season December to January, one crop in a year.

*Leaf characters.* Shape (blade) ovate-lanceolate or boradly lanceolate, size 3.5-6.2 cm. long, 1.5-2.5 cm. broad, apex sub-acute, emarginate, margin crenate; petiole broadly winged, size 3.5-7 cm. long, 1.5-26 cm. broad, shape somewhat oblong-spathulate to slightly obovate, apex obtuse, margin slightly crenate; dark green, medium coriaceous, articulated; mid-rid prominent on both sides, glabrous.

**Soh-kymphrshrieh (*C. latipes*)**

*Flower characters.* Bisexual only, clustered (2-8), occasionally solitary, cymes; pedicel 6-8 mm. long, peduncle 0.5-2 cm. long, glabrous; calyx medium coriaceous, stellate 4-lobed, deltoid, apex acute, ciliate; corolla definitely 4-merous, oval to ovate-oblong, apex attenuated, obtuse, petal light purple tinged at the margin, ephemeral, size 1.1-1.5 cm. long, 5-7 mm. broad; stamens 20-30, anther 1.5-2 mm. long, filament white, free, glabrous, 7-9 mm. long, ovary 2 mm. ht., 3 mm. diam., oblate style cylindrical, white, 6-8 mm. long including stigma, stigma large, 2-3 mm. diam; flowering season March to April.

*External fruit characters.* Shape almost globular or spherical, moderately depressed at both ends, size 4.3-5.5 cm. ht., 4.8-5.5 cm. diam, rarely 6-6.5 cm. ht., 6.6 cm. diam., weight 2½-5 oz.

(usually 3 oz.); DH index 1.06-1.13 (av. 1.08); colour light yellow or gambose; surface smooth, papillate; base obtuse, almost even, occasionally ribbed; calyx even, irregularly divided, 3-4 lobed, its obtuse; stem short; apex obtuse to slghtly truncate, smooth; areolate absent, stylar scar small, even, depressed in very small apical pit, style not persistent.

*Internal fruit characters.* Rind medium thick, thickness 5-6 mm., adherence strong; oil glands spherical to ellipsoid, small, 1 mm. long. 0.5-1 mm. broad, convex, glandular layer about 1-4 rind, aroma peculiar, distinct; mesocarp (albedo) white, spongy, texture medium soft, tasteless; axis solid, irregular in outline in medium cross section, oblique, mid. diam. 0.5-0.8 cm., segments 9-11, ahderence strong, dorsal contour usually convex to truncate; rag scanty; septa white, though and leathery; pulp-vesicles white with slight greenish tinge, texture somewhat coarse, small, shape irregular, usually fusiform, size big ones 0.7-1.4 cm. long, 2-3 mm. broad, arising only from dorsal wall of the carpel; juice scanty, ½-1 oz., sour, white; seeds many, 30-60 per fruit, shape ovoid, one side smooth, other side very slightly wrinkled, rounded end warty, micropylar end flat or slightly pointed size 1.2-1.6 cm. long, 0.9-1 cm. broad, 6-7 m. thick, testa white, tegmen light brown, chalazal spot Indian red, cotyledon white.

The botanical descriptions furnished above are built on a tree growing in the compound of the St. Edmund's College, Shillong. The fruits being inedible have little commercial value. It has been tried as a root-stock for the Khasi orange (C. *reticulata*) and is found to be incompatible.

### *Soh-shyrkhoit*

*Plant characters.* A medium to tall tree, 20-25 ft. ht., spreading; moderately dense foliage, young shoots slightly coppery tinged; thorny, 3-5 cm. long, stout; supposed to be hardy plant; one crop annualy, fruiting season December to January.

*Lead characters.* Shape (blade) broadly lanceolate, size 3.5-5.5 cm. long 2-3.2 cm. broad, size variable, apex acute to slightly obtuse, margin crenate, dark green; petiole broadly winged, 3-5 cm. long, 1.8-3.2 cm. broad, occasionally slightly larger than the blade, shape broadly spathulate, apex obtuse, margin slightly crenate, articulated.

*Flower characters.* Bisexual only, glome.ate (3-7), occasionally solitary cymose; pedicel 6-8 mm. long, peduncle 0.4-1.8 cm. long, glabrous; clayx medium coriaceous, stelate, 4-lobed deltoid, apex acute, ciliate; corolla 4-merous, very rarely 5-merous, oblong, apex attenuated, obtuse, colour light purple outside. 1.1-1.5 cm. long, 5-7 mm. broad; stamens 24-31, anther 1.5-2 mm. long, filament usually polyadelphous, some free, white glabrous, 7-9 mm. long; ovary oblate, glabrous, 2 mm. long, 3 mm. broad, style cylindrical, glabrous, white, 7-8 mm. long including stigma, stigma large, 3 mm. diam; flowering season March to April.

PLATE LXXII

**Soh-shyrkhoit (*C. latipes*)**

*External fruit characters.* Shape more or less oblate, moderately deppressed at base and apex; size 7.3-8.5 cm. ht., 8-9.8 cm. diam, weight 8-12 oz.; D/H index average 1.07; colour bright yellow; surface smooth or very finely papillate, finely pitted; base moderately depressed, minutely ribbed; apex broadly depressed;

areole absent, stylar scar even, depressed in a small apical pit, style not persistent.

*Internal fruit characters.* Rind medium thick, thickness 1.5-2.8 cm., adherence medium strong; oil glands, conspicuous, number 14-20 per 1/4 sq. cm., oil medium, aroma distinct, glandular layer about 1/4th of rind; mesocarp (albedo) chalky white, spongy, soft, almost tasteless; axis solid, mid diameter 1.5-2 cm., segments 9.15 (usually 9-12), adherence medium to strong, septa white, medium coriaceous, tough, dorsal contour convex or slightly retuse; pulp-vesicles uniformly white, texture medium coarse, soft and melting, shape usually fusiform, size 1.2-2.1 cm. long, 2-5 mm. broad, tips usually acute, arising from dorsal contour, stalked; juice 3-4 oz. per fruit, white, very sour, aroma not agreeable; seeds many, number 40-60 per fruit, shape irregular, almost orbicular to somewhat cuneate, one side smooth other side minutely wrinkled, rounded and rough, size 1.3-1.7 cm. long, 0.8-1.1 cm. broad, 5-6 mm thick, testa white tegmen light brown, chalazal cap Indian red, cotyledon white.

This variety was collected from Mawprem near Shillong. On examination of the descriptive details it will be seen that it conforms to *Citrus latipes* in basic taxonomical details but differs from the type by its bigger fruit size, greater quantity of juice and also for having flowers occasionally with five petals instead of four, characteristics of the type species. The fruits have hardly and commercial value.

## Section *Eupapeda*

### The Melanesian papeda (*Citrus macroptera* Montrouzier)

Origin and name. Swingle (1946) reported 'this species was discovered by Father Montrouzier in the Island of Art situated a few miles northwest of the north end of New Caledonia'. It is well distributed in Thailand, Indochina, Philippines, New Guinea, New Caledonia, Polynesia, etc. and thus it may be considered to be a native of those regions of south-east Asia. Its common name, as reported by Swingle, is Melanesian papeda.

The sat-kara of Assam represents the species C. macroptera Mont. and is very commonly cultivated particularly in the districts of Sylhet (now in East Pakistan), the Khasi Hills, the Mikir Hills and is also found growing gregariously in a semi-wild state in the North Cachar Hills and in the Namber-forests of Upper Assam.

It is called sat-kara (sat means seven, i.e., probably multiple of seven as satkara usually contains 14 carpels, and kara means part or carpel) in Sylhet and Cachar plains, sooh-kwit in the Khasi Hills, Hampur-arong in the Mikir Hills and Chamarbaphang in the North Cachar Hills.

Nomenclature. Citurs macroptera has been described by Montrouzier in 1860. Morphologicaly a similar fruit has also been described by Bailey (1903) under the name Citrus papuana. From a study of the original descriptions for both, it has been observed that the satkara does not fit in well with either of the congener species referred to above; but C. macroptera var. kerrii as described by Swingle (1946) bears a very close semblance to the satkara. Tanaka (1937, a) considers it to be very similar to C. macroptera var. annamensis and also to var Cambara of Rafinesque (1838). In Flora of Assam, Kanjilal, Das, and Purkayastha (1936) identified the satkara as Citurs hystrix, DC. which according to our study of the specimen does not seem to be correct because the sat-kara of Assam differs from the species C. hystrix DC. in many respects; giving rise to very tall trees upto 50 ft. and having slightly thicker and bigger leaf, flowers usually crowded (2-10), margin and apex of the petals with deep purple colouration, usually bigger sized fruits (like standard grapefruit) with almost smooth surface, 13-15 segments, plenty of juice, characteristically peculiar arrangement of pulp-vesicles arising not only from the dorsal but also from the radial walls of the carpels and somewhat thin seeds with light Indian red or sepia brown colour of the chalazal spot.

So far as our information goes it may almost definitely be stated that in Assam there exists no real Citrus hystrix, otherwise such an interesting fruit would not have escaped our noticed in course of our exhaustive collections. In the Flora of Assam the authors are obviously influenced by the work of Hooker (1872) who, it is clear, was probably confused by the leaf and fruit of the Citrus latipes whcih had been considered by him to be synonymous with C. hystrix DC. It may also be recorded here that the leaves of the fruiting terminal branches of the sat-kara are comparatively smaller in size than that of the normal and the leaves collected from such a branch are likely to be confused with those of C. hystrix. The detailed botanical descriptions of C. hystrix DC. as have been furnished by Kurz (1877) in Forest Flora of British Burma, Ochse (1931) and illustrations of the same reproduced by Bonavia (1890) and by Macmillan (1935) may be referred to for clearing up the confusion that exists between the satkara and C. hystrix.

In our considered opinion the *satkara* of Assam, as it exists to-date, in either a true representative of C. *macroptera* var. *kerrii* as described by Swingle or it is an independent variety belonging to the same species. It has, however, been placed as a variety in this work under the species. C. *macroptera*. Mont.

Another form has been isolated in Assam. This has no other morphological difference with the *satkara* except in the size and shape of fruits and to very slight extent in the size and shape of leaf.

**Keys to Varieties of Citrus Macroptera**

A. Plant very tall, 30-50 ft. ht., densely foliaged; very thorny; leaf with very large petiolar wing, often larger than blade, shape (blade) ovate elliptical; petiolar wing obovate or obcordately oblong,margin subentire; leaf including petiolar wing 9-16.5 cm. long; flowers small, usually clustered, petals white, margin and apex tinged, filaments free, somewhat wavy, pistils stud-shaped, style very short.

B. Fruit round-oblate; size 6.5-8.5 cm. ht., 7.5-10.5 cm. diam; surface almost smooth, pitted at regular intervals; segments 13-15; pulp vesicles small, obovoid charateristically arranged arising from the dorsal and radial walls of the carpels, contains oil droplet; seeds somewhat thin, cuneate, chalazal cap Indian red.

$V_1$.......*Sat kara.*

BB. Fruit ovate-pyriform; size 16-17 cm. ht., 11-12 cm. diam.; pulp-vesicles fusiform; seeds many, chalazal spot light pink.

$V_2$............*Tith-kara.*

***Sat-kara***

*Plant characters.* A very tall tree, 30-50 ft. ht., erect; very thickly branched; densely foliaged, tree with usually round top; very thorny, very stout, sharp, big ones usually 5-8 cm. long, arising almost at right angle; young branches angular, glabrous; supposed to be very handy and long-lived, very slow grower; one crop in a year, season usually November to January.

*Leaf characters.* Leaf large, coriaceous, dark green, glossy, prominently punchate with oil glands, petiolar wing usually large or larger than the blade, in old fruiting branches both blade and the petiolar wing are almost equal and comparatively smaller in size than the normal, shape (blade) ovate-elliptic, base almost obtuse, apex obtuse, emarginate, margin sub-entire or very slightly crenate; petiole broadly winged, obovate or obcordately oblong, base abruptly attenuated, apex retuse, margin almost entire to very slightly crenate, mid-rib prominent on both sides, glabrous; articulated; size blade 4.5-7.2 cm. long, 3-4.5 cm. broad with petiolar wing 4.5-9.5 cm. long, 2.5-5 cm. broad; crushed leaf aroma distinct.

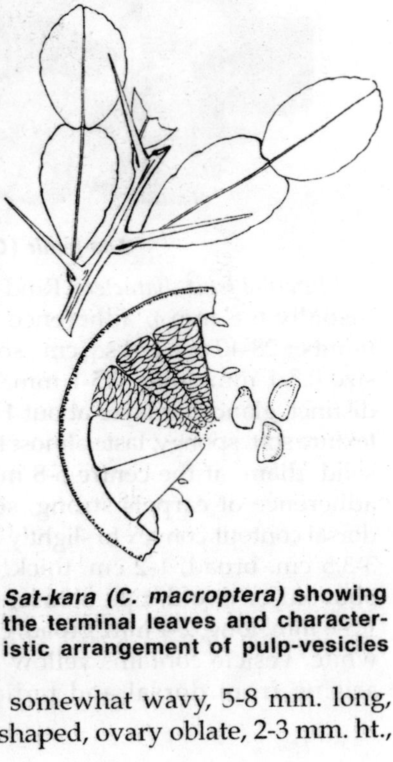

**Sat-kara (*C. macroptera*) showing the terminal leaves and characteristic arrangement of pulp-vesicles**

*Flower characters.* Bisexual only, some flowers with obortive pistil, glomerate (2-10), rarely solitary, cymose; pedicel very short, 1-2 mm. long or almost sessile. glabrous; calyx tube medium coriaceous, deltoid, apex acute, tinged; corolla imbricate, petals 4-5 white, margin and apex tinged but ephemeral, shape more or less oval, reflexed when fully opened, apex obtuse, size 0.8-1.1 cm. long, 6-8 mm. broad; stamens 23-30 (usually 27-29), filaments white, free, glabrous, somewhat wavy, 5-8 mm. long, anther 2-3 mm. long; pistil stud-shaped, ovary oblate, 2-3 mm. ht.,

4-5 mm. diam.; 13-15 locules, style very short 1 mm. long, terete, white, style fixed in a cavity of the ovary, stigma capitate, 2.5-3 mm. diam.; flowering season march.

*External fruit characters.* Shape more or less oblate to globular size 0.5-8.5 cm. ht., 7.5-10.5 cm. diam.; wt. 10-16 oz.; DH index 1.14-1.20 (av. 1.19); colour yellow to bright yellow; surface almost smooth, finely papillate, pitted at intervals, base obtuse, almost smooth or very minutely ribbed; calyx evebn, tips dried up; stem small; apex evenly rounded; areole absent, areolar area very slightly depressed or obtuse, stylar scar even or very slightly depressed in a small apical pit, style not persistent.

**Ada-jamir (*C-assamensis*)**

*Internal fruit characters.* Rind medium thick, thickness 0.6-1 cm. (usually 6-8 mm.), adherence strong; oil glands conspicuous, number 28-40 per 1/4 sq. cm., shape spherical to slightly ellipsoid, size 0.5-1 mm. long, 0.5-1 mm. broad, oil medium, aroma mild, distinct, glandular layer about 1/4 rind; mesocarp (albedo) white, texture soft, spongy, taste almost tasteless or very slightly sweet; axis solid, diam. at the centre 6-8 mm.; segments 13-15, (usually 14), adherence of carpels strong, septa medium coriaceous, tough, dorsal contour convex to slightly truncate, carpel size 5-6.5 cm. long, 3-3.5 cm. broad, 1-2 cm. thick; pulp-vesicles very small, shape obovoid or ovoid, tips obtuse, stalked, 1-4 mm. long, pulp size 1.5-6 mm. long, 2-4 mm. broad, colour white or very light greenish white, vesicle contains yellow oil droplet, arranged peculiarly, arising from dorsal and radial walls of the carpels leaving

0.8-1.1 cm. of the radial wall free towards axis; juice plenty, 3-5 oz. per fruit, very sour, colour very light greenish white to rather white, flavour agreeable; seeds many 20-40 per fruit, shape irregular, usually somewhat cuneate, some almost triangular, size 1.2-1.7 cm. long, 7-9 mm. broad, 3-5 mm. thick, micropylar end either flat or slightly pointed, surface smooth on both sides, testa cream white, tegmen very light brown, chalazal spot light Indian red, cotyledon white.

As stated before the *satkara* grows gregariously in a semi-wild condition and yields a good revenue to the Government from the sale of the fruits. The fruits are especially esteemed by the Muslims who generally preserve it for deferred use. It is often used in preparing curries and freshly killed jucice makes a refreshing cold drink with sugar. The fruits being very juicy and vesicles being very small resembling that of the lime, it is some times used for manufacturing squash and the product is labelled as lime squash.

The *satkara* has been tried as a root-stock for the Khasi orange (*Citrus reticulata*) but it does not appear so far to be promising. The stock outgrows the scion considerably and there is a definite tendency for dwarfing of the scion. On its root it is a very slow grower but attains ultimately a very good height. It is supposed to be very hardy and a long-lived plant.

### *Tith-kara*

*Plant characters.* A tall tree, 30-40 ft. ht., well branched, spreading; densely foliaged; throny, stout, 4-6 cm. long; one crop in a year, season November to December.

*Leaf characters.* Shape (lamina) ovate-oblong, size 4.5-5.5 cm. long, 2.5-3 cm. broad, apex obtuse margin slightly crenate; petiolar wing broad, shape obcordate or broadly spathulate, apex retuse, margin very minutely crenate or entire; coriaceous, dark green, mid-rib glabrous, glossy, articulated.

*Flower characters.* Solitary and glomerate (1-6), bisexual, cymose; pedicel 2-3 mm. long. peduncle usually 3 mm. long, glabrous; clayx small, 4-5 lobed, deltoid; corolla imbricate, petal 4-5, apex obtuse, tinged, size 0.6-1 cm. long, 4-6 mm. broad, obovate; stamens 24-34, anther 2-3 mm. long, filament free, wavy, 4-8 mm. long; pistil stud-shaped, style cylindrical, greenish, white, 2.5-3.5 mm. long including stigma; flowering season March to April.

*External fruit characters.* Shape more or less pyriform, or ovate-pyifrom; size usually 16-17 cm. ht., 11-12 cm. diam.; weight

1.5-2 lb.; D/H index 0.70; colour yellow; surface almost smooth, minutely pitted and papillate; basal area very slightly depressed and minutely ribbed; calyx medium coriaceous deltoid, tip obtuse; apex rounded, areole absent, stylar scar depressed in a small apical pit, style not persistent.

*Internal fruit characters.* Rind thick, thickness 1.2-1.7 cm., oil glands prominent, number 20-25 per ¼ sq. shape globular, aroma distinct; mesocarp (albedo) white, spongy, almost tasteless; axis usually solid, occasionaly semi-hollow in full ripe fruits; segments 13-14; pulp-vesicles white, not very melting, texture medium coarse, more or less fusiform and ovoid, vesicles arising from the dorsal and radial walls of the carpels; juice plenty 6-8 oz. per fruit, very sour; seeds many, 25-38 per fruit, shape cuneate or somewhat triangular, size usually 1.5-1.6 cm. long, 0.9-1 cm. broad, 4-6 mm. thick, pointed or slightly flat at the micropylar end, surface smooth on both sides, ridged on the edges, testa white, tegmen lgiht brown, chalazal spot light pink, cotyledon white.

This variety is occasionally met with in the district of Cachar and the specimen, detailed above, has been collected from Karimaganj.

# Chapter 4

# New Species

## *Citrus assamensis* Dutta and Bhattacharya. Sp. Nov.

*Introduction.* The specimen of *ada-jamir,* as it is known locally, has been collected from an interior village in Karimganj sub-division of the district of Cachar. Identical specimens have also been found to occur particularly in Sylhet (now in East Pakistan), North Cachar Hills and Khasi Hills. It is sporadically grown in the homestead gardens. The fruits are valued locally for its peculiar aromatic flavour and intense sour juice. The aroma of the rind approaches to eucalyptus smell but people characterise it to be similar to that of the ginger (*Zingiber officianale*) and hence the name *ada-jamir* (*ada*=ginger; *jamir*=citrus). It is also called as *soh-sying* (*soh=fruit; Sying*=ginger) in the Khasi Hills of Assam.

*Nomenclature.* The material has been examined in fresh condition from taxonomic points of view. It has neither been found to fit in with any of the well recognized species or varieties of the genus *Citrus* nor can it be suspected to have originated as an accidental hybrid. The original specimen had however, been sent to Kew Herbarium (England) for identification. The memorandum of identification suggests that it has some close alliance with *C. hystrix* DC. in respect of certain characters of leaf and androecium. It has been further stated that 'There are of course other characters that keep the Assam plant distinct from the typical, *C. hystirx*' and it concludes that in absence of any intensive knowledge of the group, it should be considered as a variety of *C. hystrix* DC. From the above statement it will appear that in the matter of identification of the *ada-jamir* of Assam the memorandum rather leans on assumption. The writers, however, are unable to subscribe to the above view. The descriptions of *C. hystrix* DC. as built up by Rumphius (1841), Kurz (1877), Ridley (1922), Ochse (1931), Swingle (1946), and others have been very carefully compared with those of the *ada-jamir.* Besides Tanaka (1937, c) categorically points out that *C. hystrix* does not occur in India. This lends a further prop to our contention. It may also be stated that living plants of *C. hystrix*, obtained from Ceylon, have also been grown and examined side by side with the *ada-jamir* at the Citrus Fruit research Station, Burnihat. The studies reveal that the two specimens differ wide apart. The description of *C. hystirx* DC. or *Lemo-agrestis* of Rumphius which has been considered by Merril (1917) to be identical with *C. hystrix* DC., bear only certain amount of superficial similarity with that of Assam plant; but differs from it fundamently in important diagnostic details. The *ada-jamir* of Assam, though coming near *C. hystrix* DC., differs from the latter in having :

(1) Medium sized moderately branched tree;

(2) Large leaf blades 7.5-20 cm. long, 3.5-10 cm. broad, crenate to bi-crenate with revolute margin, prominently marked with pellucid glands both in marginal notch and in surface,leaf aroma distinct, strong, approaching to eucalyptus smell; petiolar wing with distincly revolute margin, obovate to spathulate in shape and never obcordate, petiole usually about 1/3 both in length and width of the blade and never equal or larger than the blade;

(3) Periodical habit of flowering, glomerate (4-9) and never solitary, medium to large flowers and never small, peals deep purple colour and never white, 4-8 merous, stamens 26-42, filament polyadelphous and never all free, vase-shaped ovary;

(4) Fruit spherical to slightly turbinate with almost smooth surface, pushed up mamilla with obtuse to truncate apex and never depressed, size medium, 7-9.5 cm. ht., 7-9.8 cm. diam., segments 10-15, abundant white juice, vesicles entirely free form acrid oil droplets, fusiform, and never round or oval (presence of round or oval shaped pulp-vesicles with acrid oil droplets is characteristic of *C. hystrix*), juice not bitter; seed slightly smaller than that of *C. hystrix*, ovate or D-shaped, cotyledon light greenish white and never pure white.

In view of all these basic differences between the two there cannot be any valid ground for grouping them together under the species *C. hystrix* DC. and in the fitness of things the *ada-jamir* of

Assam deserves a separate specific rank other than C. *hystrix*. It may also be stated here that no reference to *ada-jamir* of Assam deserves a separate rank other than C. *hystrix*. It may also be stated here that no reference to *adajamir* of Assam deserves a separate rank other than C. *hystrix*. It may also be stated here that no reference to *ada-jamir* of Assam has been made by Roxburgh (1832), Hooker (1872), Brandis (1874, 1906), the early explorers of the Indian flora, in their classical works. Bonavia (1888-90) too had not mentioned or illustrated the *ada-jamir* though he dealt with various *citrus* of Assam. Tanaka visited India several times from 1928-1937 and described many new Citrus species and varieties of Assam but in his exhaustive list of Citrus varieties he did not mention even once about *ada-jamir* or *sohsying* so very common in the Khasi Hills of Assam which he had actually visited. Kanjilal, Das and Purkayastha (1936) in their *Flora* of *Assam* did not also refer to the *ada-jamir*. Thus it appears that it has, hitherto, been described by none. This has now been described for the first time by the present authors who propose to raise it to the rank of a new species by calling it as *Citrus assamensis*.

## KEY TO CITRUS *Assamensis* sp. nov.

A. Pulp-vesicles free from acrid oil droplets, intensely sour and not bitter.

B. Plant medium-sized; much thorny, stout; leaf glossy, coriaceous ovate-oblong; moderately broad petiolar wing, slightly obovate to spathulate, about 1/3 of length and breadth of lamina, margin bicrenate, definitely revolute; crushed leaf aroma approaching to eucalyptus small; flowers more than once annually, medium large glomerate, (never solitary), petals 4-8, deep purple, stamens 26-42 polyadelphous, some free; ovary vase-shaped, 10-15 locules.

C. Fruit medium-sized (usually 7-.5 cm. diam), almost spherical; surface almost smooth; apex obtuse, pushed up mamilla; axis hollow; pulp fusiform, slender, (never oval or globose); seeds many, D-shaped, chalazal spot Indian red, cotyledon light greenish white.

C. *assamensis* (Sp. Nov).

## *C. assamensis Dutta* and *Bhattacharya. Sp. Nov.*

**Citrus assamensis sp. nov. Green natured Fruit**

*Planta* est altituoline media, valde spinosa, spinis magnis es acutis; *folia* sunt larga, coriacia; magnopere splendent; ovato-oblonga; maculis punctuata pellucidis; margine bi-crenato, petiolum largitate media, tertia circiter pars folii, figura spathulato-obovata, margine curvo crenato, pellucidis maculis in unaquoeque incisura, folium cum atteritur, exsurgit aroma forte, simile eucalypto; *floret* plus quam semel in anno, flores sunt largitate media, bi-sexual cum staminate, gllomerati (4-9), cymosi, 4-8 fere petalia, superficies cotoris purpurei, stamina 26-42, filamina juncta 2-5 vel bibera, glabra, anther, longtituline 3-5 mm., si stigma comprehenditure, viridi albus color mutatur stylus glabrus, 7-9 mm., et fit purpureus, stigmalis diameter 4.5 mm., ovare figura vasculum, 10-15 loculare. *Fructus* fare rotundus, aliquantulum turbinatus; figura generatim 7.0-9.5 cm. longo, 7.0-9.8 cm. diameter, color citron parum viridescentis, superfices fare plana interdum verrucis parvis iniquo; ad basim collum minimum et leviter depressus costis praeditus; apex valde leviter verrrucuatus vel inverrucam ejectus, areole cum nota circularis irregulari; minutatim costis randiantibus signita etsi non semper eodem modo, cortex 5-7 mm., glandulis oleiferis conspicuus; figura ovali elliptica; aroma forte sicut eucalypti; axis excavata, interdum solida; segment 10-15; pulpal vesicuale albae, testura pura, mollis, figura fusiformis, exteremitates generaliter praeacutae, saporosae, valde amarae; semina multa, 1-1.2 cm. × 6-8 mm., × 4-5 mm., figura fere ovali aliquanolo D-forma, superficies prope plana, una extereminate minutatim rugatae, testa alba in modum lactis, tegmen leviter fuscum, chalazal ruber, cotyledon viridiallium.

### *C. assamensis Sp. Nov.*

*Plant characters.* A medium to tall plant, 15-25 ft. ht., branching low from the trunk moderately branched, spreading; moderately dense foliage; very much thorny, stout, very long, 0.5-7 cm. long, right-angled; young branches angular, glabrous, glossy, young shoots tinged; frutiing season mainly in September to November, minor crops in other parts of the year; plant very much susceptible to gummosis and canker.

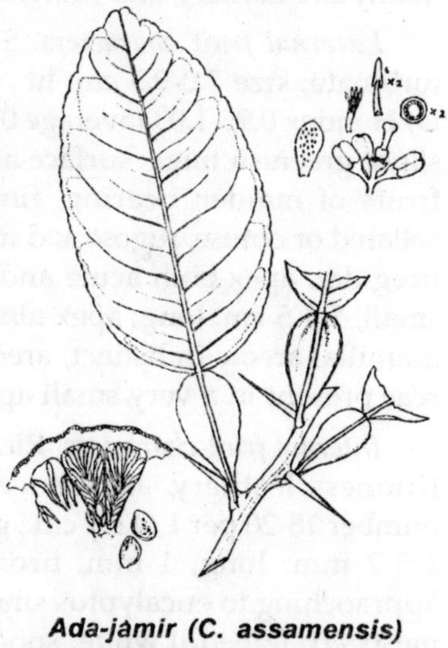

**Ada-jamir (C. assamensis) showing terminal leaves and thorns and internal structure of fruit and flower**

*Leaf characters.* Leaf large with moderately broad petiolar wing; petiole usually about 1/3 both in length and in width of the blade size (blade) 7.5-20 cm. long, 3.5-8.5 cm. broad with petiolar wing 2.5-6.5 cm. long, 0.9-3.7 cm. broad, shape (blade) ovate-oblong coriaceous, deep green, very glossy, margin revolute, bi-crenate, to crenate, prominently punctate with oil glands in every notch, apex obtuse, emarginate; petiolar wing slightly obovate to spathulate, margin minutely crenate, dotted with oil glands,definitely revolute, apex abtuse, articulated; crushed leaf aroma strong approaching to eucalyptus smell, distinct of the type.

*Flower characters.* Bisexual and staminate, glomerate (4-9), never solitary, cymes; pedicle short, 5-7 mm. long, peduncle 0.6-2 cm. long, glabrous; calyx more or less cupular, greenish white, tips purple tinged, apex acute, deltoid, ciliate, 4-5 lobed; corolla imbricate, petals 4-8, shape oblog, tip obtuse, colour deep purple outside, white inside, size (petals) in male flower 1.7-2 cm. long, 7 mm.-1 cm. broad in bisexual flower 2-2.1 cm. long, 9 mm. -1.2 cm. broad; stamens 26-42, filaments white, glabrous, jointed by 2-5 together, polyadelphous, some free (8-12), 0.8-1 cm. long more or less triangular in cross section, anther 3-5 mm. long; ovary 4-5 mm. long, 4-5 mm. broad, somewhat vase-shaped or globose, 10-15 locules, style terete, glabrous, 7-9 mm. long, greenish white, turns into purple, stigma diameter 4-5 mm.; periodical habit of flowering, mainy in February and March.

*External fruit characters.* Shape almost spherical to slightly turbinate; size 7.0-9.5 cm. ht., 7-9.8 cm. diam.; weight 10-14 oz., D/H index 0.95-1.03 (average 0.98); colour light lemon yellow with slight greenish tinge; surface almot smooth, very slighty warty in fruits of maiden bearing, finely foveolated; base usually low collared or obtuse, rugose and minutely ribbed; calyx small, division irregular, apex both acute and obtuse, medium coriaceous; stem small, 1-1.5 cm. long; apex almost obtuse to truncate, pushed up mamilla; areole indistinct, areolar area rugose and ribbed, stylar scar present in a very small apical pit, style not persistent.

*Internal fruit characters.* Rind medium thick, thicknes 5-7 mm. firmness leathery, adherennce strong; oil glands conspicuous, number 18-20 per 1/4 sq. cm., glands usually oval or ellipsoid, size 1.5-2 mm. long, 1 mm. broad, oil abundant, aroma strong, appraoching to eucalyptus smell, glandular layer about 1/2 rind; mesocarp (albedo) white, spongy, soft, sweet; axis hollow, rarely solid, diameter at the centre 1-1.5 cm.; segments 10-15, adherence strong; rag little; septa chartaceous tough, white, dorsal contour convex to truncate carpel broken or closed at axis; pulp-vesicles white in matured fruit, free from acrid ol droplets, texture fine, slender, soft and melting, fusiform stalk, arising from dorsal wall and the lower portion of the septa, tips usually actue; flavour excellent; juice abundant, colour white, weight 4-7 oz. per fruit, very sour, not bitter seeds many, 30-40 per fruit, size 1-1.2 cm. long, 6-8 mm. broad, 4-5 mm. thick, shape almost ovate to somewhat d-shaped, micropylar end almost obtuse, slightly flat, surface almost smooth, convex, one edge minutely ridged, testa cream white, tegmen very light brown, chalazal spot Indian red, cotyledon very light greenish white.

# Chapter 5

# PROBABLE HYBRIDS

In course of collection of different Citrus materials from various parts of the province a number of specimens have been garnered, which have not found to fit in adequately with any of the recognized species. A detailed study of each of these apparently 'off types' was undertaken to see whether any speicfic or vareital rank can be assigned to them. The taxonomic and observational studies of the materials strongly indicate the these are nothing but chance hybrids, originated probably by natural cross-pollination whch is not infrequent in the genus *Citrus.* The parentage of these accidental hybrids is very difficult to ascertain at the present stage of our knowledge but some of the dominant characters as expressed in them bear striking resemblance with some of the prominent varietal characters of known recognized species with which these are probably genetically linked. The citation of probable parentage for these accidental hybrids is at best suggestive and tentative which can be settled definitely only if safe guarded hybridization is carried out.

The propriety of incormporating descriptions of these natural hybrids in the present monograph lies in their horticultural importance as an omission would virtually mean a negation to their very existence. It is undeniable that some of these hybrids, particularly *soh-khyllah* (No.1), *Nichols-Roy,* hold considerable promise for commercial exploitation. It is therefore, considered expedient, that these should also find a mention at least along with other well recognized varieties.

### *Soh-kyllah No. 1*

*Plant characters.* A tall tree, 25-30 ft. ht., clean trunk erect; densely foliaged; usually thornless; fruiting season December to February, one crop annually.

**_Soh-khyllah_ (hybrid No. 1)**

*Leaf characters.* Leaf ovate-lanceolate, deep green, margin crenate, apex acute, emarginate, size 6-13 cm. long, 3-7.5 cm. broad, with petiolar wing 1.5-3 cm. long, 3.8 mm. broad; petiolar wing spathulate, apex obtuse, margin usually entire, articulated.

*Flower characters.* Bisexual only, cymes; pedicel 4-6 mm. long, peduncle 0.6-2.5 cm. long, glabrous; calyx deltoid, apex acute, ciliate, greenish white, 4-5 lobed; corolla imbricate, 4-5 petals, pure white; oblong, size 1.5-1.8 cm. long, 6-8 mm. broad; stamens 19-20, filament polyadelphous, jointed 3-6 together, glabrous, white, 1-1.3 cm. long anther 2-3 mm. long; ovary almost globose, smooth, size 3 mm. long, 3 mm. broad, style cylndrical, glabrous, greenish white, 8-9, mm. long including stigma, stigma capitate, 2-2.5 mm. diam; flowering season March to April.

*External fruit characters.* Shape turbinate to ovate pyriform, drawn-out neck; size medium, 8.5-10 cm. ht., 8-9.5 cm. diam., rarely 12.5-13 cm. ht., 10-10.7 diam; weight usually 10-14 oz.; D/H index 0.81-0.97 (av. 0.90); colour bright yellow to orange yellow; surface almost smooth to rugose, finely pitted; base very low-collared; apex slightly depressed, naval absent; areole indistinct, areolar area slightly rough and radially ribbed, stylar scar present even in a depressed pit, style not persistent.

**_Soh-khyllah_ (hybrid No. 1) showing internal structure of fruit and outline of leaf.**

*Internal fruit characters.* Rind medium thick, thickness 0.6 cm. (usually 6-8 mm.), firmness soft, adherence slight to medium; oil glands many, conspicuous, number 20-30 per ¼ sq. cm.,shape pyriform and ellipsoid, size medium to large 2-5 mm. long, 1-3 mm. broad oil abundant, aroma weak, pleasant, glandular layer ½ rind; mesocarp (albedo) white, soft and spongy, taste sweet, vascular bundle prominent; axis semi-hollow or solid, diameter at the centre 0.8-1.5 cm., segments 11-16 (usually 11), adherence medium, septa medium coriaceous, tough, dorsal contour truncate to slightly convex or somewhat wavy, carpel size 6-8.3 cm. long, 2.5-3.5 cm. broad, 1-2.8 cm. thick; pulp-vesicles colour honey yellow, texture coarse-grained, shape big ones fusiform, others polygonal, size 1.5-2.4 cm. long 2-4 mm. broad, tip obtuse, arising from dorsal side of the carpellary wall, stalked, flavour agreeable; arising from dorsal side of the carpellary wall, stalked, flavour agreeable; juice plenty, 4-7 oz., sweetness and acidity blended, colour honey yellow; seed many, 20-24 per fruit, size 1.6-2.2 cm. long, 8 mm.-1 cm. broad, 5-7 mm. thick, somewhat cuneate, micropylar end beaked,one side with some smooth area and other with anastomosing lines, testa cream white, tegmen light brown, chalazal cap Indian red, cotyledon usually white to light greenish white.

This is suspected to be a hybrid, originated by natural crossing between the species *C. reticulata* and *C. grandis*. The Khasi name soh-khyllah (*soh*=fruit; *khylla* = strange or uncommon) means a strange or uncommon fruit. The very name thus suggests that it is of hybrid origin. It bears a somewhat close semblance to Sampson Tangelo as described by Webber (1946) in the *citrus Industry* Vol 1., pp. 647. The *soh-khyllah* No. 1 represents a good type of tangelo of Assam. The fruits are lined by many and valued in market. This is grown as novelty in home gardens in the Khasi Hills,

### *Soh-khyllah No. 2*

**Soh-khyllah (hybrid No. 2)**

*Plant characters.* A medium sized tree, 20-25 ft. ht., spreading moderately dense foliage, erect habit; usually almost thornless; young branches angular, glabrous; fruiting season December to January, one crop in a year.

*Leaf characters.* Lead medium coriaceous, deep green, shape usually broadly lanceolate, creante, apex obtuse, emarginate, size 7.5-12 cm. long, 3.4-5.7 cm. broad, wtih petioles 1.5-2.5 cm. long, 2-6 mm. broad, articulated; veins slightly prominent on both sides.

*Flower characters.* Bisexual, cymose, solitary and clustered (1-8); pedicel 3-4 mm. long, peduncle 2-6 mm. long, glabrous; calyx more or less stallate, 4-5 lobed, deltoid, apex ciliate; corolla imbricate, petal 4-5, pure white, oblong, size 1.2-1.6 cm. long, 5-8 mm. broad; stamens 22-25, filament usually free, white glabrous, 0.7-1.1 cm. long, anther 2-2.5 mm. long; ovary usually 2 mm. ht., 2 mm. diam., style greenish white, cylindrical, glabrous, 6-7 mm. long including stigma, stigma capitate 2 mm. diam.; flowering season March to April.

*External fruit characters.* Shape almsot spherical to oblate bumpy, size 10.0-13.5 cm. ht., 11-14.5 cm. diam.; weight 14-25 oz.; D/H index 1.05-1.08 (av. 1.06); colour orange; surface warty, undulating, uniformly foveolated; base deeply depressed, ribbed, basal area very warty; calyx medium thick; stem usually 2.2-3 cm. long; apex broadly depressed; areole indistinct, stylar scar present, small, even, style not persistent.

*Internal fruit characters.* Rind somewhat thick, thickness 0.7-1.1 cm. (usually 8 mm.), firmness almost soft, adherence very slight or none; oil glands medium to large, size variable, 3-5 mm. long, 1-2 mm. broad, pyriform or elliptical or globose, number 28-36 per ¼ sq. cm., oil abundant, aroma distinct, glandular layer ¾ of rind; mesocarp (albedo) white, spongy, almost tasteless or with a soupcon of sweetness, vascular bundle prominent reticulated over the carpels; axis hollow, diameter at base 1-1.5 cm., at the centre 3-3.5 cm., at apex 2.5-3 cm.; segments 12-14, adherence slight, septa medium coriaceous, tough, dorsal contour more or less wavy, carpel size 6-7 cm. long, 3-3.5 cm. broad, 1.5-2.5 cm. thick; pulp-vescicles light orange tinged, texture coarse, melting, arising from dorsal and carpellary walls, shape irregular, big ones usually fusiform, polygonal, size big ones 1-1.4 cm. long, 4-7 mm. broad, tip obtuse, flavour distinct; juice abundant 6-8 oz. per fruit, colour light orange, taste sweetness and acidity blended rather acidulously sweet; seeds number 15-20 per fruit, size 1.4-2 cm. long, 0.9-1 cm. broad, 5-6 mm. thick, shape somewhat cuneate, pointed or flat towards micropylar end, rounded end usually ridged, testa cream white, tegmen light brown, chalazal cap deep brown, cotyledon very light greenish white.

The fruit sems to be a complex hybrid, orignated probably from *C. reticulata, C. grandis* and *C. aurantium*. It almost resembles pummelo in size, *Citrus aurantium* in size and shape of seed and aroma of rind, and *C. reticuulta* in shape and size of leaf, its separable rind and carpels, colour and shape of vesicles and colour of cotyledon. The fruit on the first look appears to be an 'elephant' form of mandarin. Similar form of fruits has been pictured in Plate Nos. CX to CXIII by Bonavia (1890) under the name *at-ani* which indicates an orange half the size of pummelo. The *soh-khyllah* No. 2 of Assam undoubtedly represents the *at-ani* of Bonavia. Lushington (1910) considers the *at-ani* as a variety under the mandarin orange and calls it as *C. chrysocarpa* var. *decumana*, while Tanaka (1937, a) suggests a separate specific name *Citrus rugulosa* Hort. for the same.

## *Soh-khyllah No. 3*

**Soh-khyllah No. 3**

*Plant characters.* A tall tree, 25-30 ft. ht., erect habit, densely foliaged; thorny, 3-3.5 cm. long; fruiting season December to February, one crop in a year.

*Leaf characters.* Leaf medium coriaceous, dark green, shape ovate oblong, size 4.8-10 cm. long, 2.7-6.0 cm. broad with redimentary petioles, 9.8-2.3 cm. long, 2-5 mm. broad, margin (blade) crenate, apex obtuse, occasionally slightly bifid, articulated.

*Flower characters.* Flowers bisexual only, solitary and clustered, (1-6), cymes; calyx star shaped, deltoid, apex acute, ciliate, greenish white; corolla imbricate, petals 4-5, pure white, oblong, size 1.2-1.6 cm. long, 5-8 mm. broad; stamens 17-20, filament white, usually polyadelphous, glabrous, 0.7-1 cm. long, anther 2-3 mm. long; ovary usually 2 mm. long, 2 mm. broad, style cylindrical glabrous, greenish whtie, 6-7 mm. long including stigma, stigma capitate, 2 mm. diameter.

*External fruit characters.* Shape spherical to slightly oblate; surface usually smooth, glossy, minutely pitted; colour golden orange to light orange, size 5.6-6.8 cm. ht., 6-7 cm. diam; weight 5½-7½ oz.; D/H index 1.03-1.16 (av. 1.10) base obtuse, minutely ribbed or rugose; calyx even deltoid, 4-5 lobed; stem 5-8 mm. long; apex obtuse to slightly truncate with a mild depression, navel absent; areole absent, areolar area even, surface almost smooth to very slightly rugose; stylar scar present, small, diameter 1-2.5 mm., slightly depressed in a very small apical pit.

*Internal fruit characters.* Rind thick, thickness 3-4 mm., firmness leathery adherence medium; oil glands many, 60-70 per sq., cm., size 1-2 mm. long, 1-1.2 mm. broad, shape big ones somewhat pyriform, small ones globose near the periphery, oil medium, aroma somewhat spicy, peculiarly distinct, glandular layer slightly more than ½ rind; mesocarp (albedo) thin, white spongy, texture soft, taste sweet; axis semi-hollow to hollow, mid-diameter about 1-1.3 cm.; segments 9-10 (usually 10), adherence medium to slight, carpel more or less easily separable, septa thin, white with very slight orange tinge, dorsal contour convex to slightly truncate, carpel size 4.7-5.5 cm. long, 2-2.9 cm. broad, 1.5-2 cm. thick; pulp-vesicles uniformly light orange, texture somewhat coarse, not very melting, shape irregular, usually fusiform, size big ones 1.2-1.6 cm. long, 2-4 mm. broad; juice medium, weight, 2½-4 oz. per fruit, colour light orange, taste sweet and acidity more or less blended, rather acidulously sweet; seeds many, 20-26 per fruit, shape more or less cuneate, surface rough with anastomosing lines, micropylar end flat, teta white, tegmen light brown, chalazal spot Indian red, cotyledon light green.

The *soh-khyllah* No. 3 is almost similar to the so called Tample orange of repute. This is probably an accidental hybrid, originated by natural crossing between the Khasi orange (*C.reticulata*) and the *soh-niangriang* (*C. sinensis*) of the khasi Hills. Though this fruit bears a strong similarity with the Tample orange there are some characters which distinguish one from the other. The fruit is decidedly inferior to the Tample orange in quality. This type of fruit is not very common and is generally planted in home-stead garden for curiosity. It is the tangor class of fruit of Assam. The original specimen has been collected from the Khasi Hills.

## *Hash-khuli*

*Plant characters.* A medium to tall tree, 18-20 ft. ht., slightly lodging habit; spreading branches, moderately dense foliage; almost thornless; young shoots slightly purple tinged, pubescent; two crops in a year.

*Leaf characters.* Leaf medium to large with broad petiolar wing, coriaceous, deep green, size big 8-22 cm. long, 5-17 cm. broad with petiolar wing, 2-6.3 cm. long, 1.2-6 cm. broad, shape (blade) ovate-oblong, margin crenate, slightly revolute, apex usually obtuse, bifid; petiolar wing obcordate, margin crenate, revolute, apex retuse; mid-rib pubescent; articulated; crushed leaf aroma resembles to that of *C. assamensis.*

*Flower characters.* Flower clustered (6-10), large, bi-sexual only, cymes; pedicel 1-1.7 cm. long. peduncle 1.5-4 cm. long, pubescent; calyx coriaceous, cupular, tips purple; corolla imbricate, petals 4-5, oblong, very slightly purple coloured, ephemeral, size 2.2-2.6 cm long, 0.8-1.1 cm. broad; stamens 22-26, filament white, some free and some jointed at the base, glabrous, 1-1.3 cm. long, anther 3-4 mm. long; style cylindrical, 1-1.1 cm. long including stigma, write, turns into light purple; ovary vasculum, size 5-6.5 mm. long, 4-6 mm. broad; flowering season February and March, second bloom in September and October.

*External fruit characters.* Shape turbinate to oblong-ovate-pyriform, size 8-19.5 cm. ht., 8.5-15.2 cm. diam.; D/H index 0.83-0.96 (av. 0.90); colour yellow; surface smooth, finely pitted, occasionally with minute ridges over the surface running from base to apex; calyx depressed in a cavity, sepals medium thick; stem 1.5-1.8 cm. long; base slightly necked or low collared and depressed, occasionally ribbed; apex obtuse, depressed, occasionally radially ribbed, navel absent, areole absent, stylar scar present, depressed in a small apical pit, style not persistent.

*Internal fruit characters.* Rind thick, thickness 0.8-2 cm.; oil glands conspicuous, number 8-12 per ¼ sq. cm., shape usually oval, size 1.5-3 mm. long, 1 mm. broad, oil little to medium, aroma distinct, glandular layer ⅕ or ⅙ th of rind; mesocarp (albedo) white, spongy, taste slight bitter with soupcon of sweetness; axis hollow,diameter at the centre 2-3.5 in certain fruits particularly in small and medium ones core is almost filled up with the protruding tips of the vesicles; segments 10-12 (usually 10-11), adherence strong, septa medium coriaceous, tough and leathery, dorsal contour truncate to retuse, occasionally convex to wavy,; pulp-vesicles white, texture medium coarse to fine, slender, medium soft, shape usually fusiform, to acicular, sometimes torulose, tips usually acute, size big ones 1.6-3 cm. long, 2-4 mm. broad, arising from dorsal side of the carpel, stalked; juice plenty, 6-10 oz., taste sour, flavour not good, seedless.

This is suspected to be a hybrid between *C. grandis* and *C. assamensis* (*ada-jamir*) described in the text. The fruit bears similarity with the pummelo in its size and bitterness of its albedo, while the oil glands with aroma of the flavedo resemble those of *C. assamensis*.

The fruit is worthless and has hardly any demand in the market. The specimen has been collected from Karimganj.

### *Sarbati*

*Plant characters.* A medium sized plant, 18-20 ft. ht., moderately branched and foliaged; almost thornless; fruiting season December to January, one crop annually, prolific fruiter.

*Leaf characters.* Medium-sized leaf with narrow wing; shape broadly lanceolate, somewhat coriaceous, light green, margin crenate apex obtuse to acute, occasionally bifid; petiolar wing (shape) usually obcordate to spathulate, margin crenate in the broad wing, more or less entire in spathulate wing, apex obtuse to truncate; mid-rib glabrous; size (blade) 9-14.5 cm. long, 4.5-8.5 cm. broad with petiolar wing 1.2-2.4 cm. long, 0.4-2 cm. broad.

*Flower characters.* Bisexual and staminate, proportion of bisexual flower more than the staminate ones, crowded (2-12), occasionally solitary, cymes; pedicel 4-9 mm. long, peduncle short, 0.5-1.6 cm. long, pubescent; calyx more or less cupular, margin tinged; corolla imbricate, petals 4-5, colour purple outside, oblong, size 1.7-2.6 cm. long, 5-9 mm. broad; stamens 30-39, filament white, glabrous, 0.7-1.3 cm. long, anther 3-4 mm. long; ovary slightly oval, size usually 4 mm. ht., 3 mm. diam., style greenish white, purple on maturity cylindrical, 1-1.1 cm. long including stigma; Stigma capitate, diameter 3-4 mm., flowering season March.

**Nichols-Roy**

*External fruit characters.* Shape glabular to occasionally somewhat turbinate, size 6.5-12.1 cm. ht., 6.5-11.5 cm. diam.; weight 5½-16½ oz. (usually 11-14 oz.; D/H index 1.00-1.11 (av. 1.04); colour lemon yellow; surface almost smooth to very slightly rough, very minutely pitted, dull; base almost obtuse or very low collared and depressed, minutely ribed; calyx even or depressed, tips pointed; stem small, usually 1 cm. long; apex almost obtuse to truncate; areole indistinct, irregular circular furrow, radially ribbed, diameter of areolar area 1.5-2 cm., style persistent, 0.5-12 cm. long.

*Internal fruit characters.* Rind medium thick, thickness 0.6-1.2 cm. (usually 6-8 mm.), firmness very firm, adherence strong; oil glands conspicuous. number 15-35 per ¼ sq. cm., shape usually globular to slightly ellipsoid, occasionally necked, size 0.5-3 mm. ht., 0.5-2 mm. diam., oil little, glandular layer ⅓ rind; mesocarp (albedo) white, texture hard, carrot like consistence, taste sweet-bitter; axis usually solid, occasionally semi-hollow, diameter at the centre 5-8 mm.; segments 10-11, adherence strong, septa medium thick, tough and leathery, taste of septa bitter, dorsal contour convex to truncate; pulp-vesicles colour uniformly white, texture medium coarse, size big ones 1.5-2.4 cm. long, 2-4 mm. broad, shape usually fusiform, not so melting, arising from dorsal side of the carpel, tips, either acute or obtuse; rag somewhat coarse; juice medium, 1½-3½ oz., white, taste rather more sour than sweet; seeds few to many, 8-30 per fruit, medium-sized, 1.2-1.4 cm. long, 6-9 mm. broad, 4-5 mm. thick, shape ovate, somewhat plano-convex, micropylar end either wedged or acute, surface one side with anastomosing lines, other side smooth, testa cream white, tegmen light brown, chalazal spot Indian red, cotyledon whtie.

***Sarbati* (natural hybrid No. 5) showing outline of leaf and internal structure of fruit.**

This fruit has been collected under the name *sarbati* which generally stands for sweet lime (*C.limetta*). The study of the morphological characters of plant, leaf, flower and fruit reveals that it is a probable hybrid, originated by natural cross pollination between the species C. *medica* (Citron) and C. *aurantium* (Bitter orange). The shape and size of leaves and fruits and taste of albedo are indicative of association with Seville or Sour orange, while habit of plant, colour of the flower, texture of rind (carrot-like consistence) and persistent style are suggestive of Citron.

It is cultivated to a very limited extent in home-stead gardens for occasional use. The specimen has been collected from Karimganj. The fruit is worthless and does not hold out any promise.

## *Nichols-Roy*

*Plant characters.* A tall free, 25-30 ft. ht., trunk diam. about 20 inches, erect habit, not very thorny, spreading, densely foliaged; leaf oblong to almost broad lanceolate, medium coriaceous, dark green, petiole narrowly winged or with rudimentary wing, size 8-12.5 cm. long 4-6.5 cm. broad with petiole 1.5-2.5 mm. long, 2-7 mm. broad; blade margin crenate, apex usually acute or slightly obtuse; articulated; mid-rib glabrous; flowers large, petals 4-5, pure white, stamens 24-32, filaments white, glabrous, polyadelphous; ovary oval, 10-14 locules; flowering season March to April; prolific fruiter, one crop annually.

**Nichols-Roy**

*External fruit characters.* Shape somewhat ellipsoid to almost spherical, size usually 11.5-12.5 cm. ht., 10.5-11.5 cm. diam. wt. 20-28 oz., D/H index average 1.08; surface almost smooth, very finely papillate, glossy, colour deep yellow to slightly tawny-yellow; base obtuse, very rarely slightly ribbed calyx medium thick, irregularly divided, deltoid, tips usually dried up; apex obtuse, areolar mark absent, very slightly depressed, stylar scar even in a small apical pit, style not persistent.

*Internal fruit characters.* Rind medium thick, thickness 5-8 mm., firmness medium soft to soft, adherence medium strong; oil glands number 12-18 per ¼ sq. cm., primary glands usually ellipsoid, size 2-3 mm. long. 1-1.5 mm. broad surface convex, oil medium, aroma mild, pleasant, glandular layer about ⅓ rind; mesocarp (albedo) white, medium thick, taste bitter; axis semi-hollow, mid-diameter 1-1.5 cm.; segments 10-14 (usually 12), adherence of carpels medium; rag medium septa slightly coriaceous, white, tough, dorsal contorn convex to truncate, carpels not broken at the core; pulp-vesicies white; texture medium coarse, tender, very soft and melting, arising from the dorsal walls, shape usually fusiform; juice abundant, wt. 10-15 oz., colour clear white, sour (mild), not bitter, flavour good; seeds many, 50-60 per fruit, size large, 1.8-2.1 cm. long, 0.8-1 cm. broad, 5-6 mm. thick, shape usually cuneate rounded end rough, one side marked with anastomosing lines, other side with a very small smooth area, micropylar end flat and beaked, testa white, tegmen very light brown, chalazal spot Indian red, cotyledon white.

The variety has been collected from the gardens of Rev. J.J.M. Nichols Roy a pioneer orange grower of Burnihat, Assam. It is named as *Nichols-Roy* after the name of the owner, who planted the original tree sometimes in 1926. It is stated that the variety has originated as a chance seedling and is supposed to be an accidental hybrid of the pummelo (*C. grandis*) crossed with some types of lemons (*C. limon*). The fruit compares favourably with the grapefruit though lacks bitter principle with which the latter is endowed. The fruit is very attractive and delicate and requires careful handling in plucking and packing. The variety though sour (mild) in taste is very juicy and mixed with sugar it makes a very refreshing compote. It holds out a coniderable promise as a variety under grapefruits in suitable locality. In order to establish a clonal horticultural variety, it has been propagated by vegetative means at the Citrus Fruit Research Station, Burnihat.

### *Dewa-tenga*

*Plant charactes.* Pant medium to tall, 15-20 ft. ht., erect; densely foliaged, well branched, spreading; yong parts pubescent;almost thornless; leaves medium large to large with broad petiolar wing, leaf blade 10-18 cm. long, 6-10 cm. broad with petiolar wing 3.5-6 cm. long, 2.2-5 cm. broad, shape (blade) broadly lanceolate, coriaceous, deep green, glossy, articulated, margin crenate to slightly serrate, occasionally entire, particularly lower half of the blade, apex almost obtuse bifid; wing (shape) obcordate, margin usually entire; veins slightly prominent on both sides; crushed leaf aroma distinctly like that of the Seville orange; flower bisexual, large, crowded, raceme, colour pure white.

*External fruit characters.* Shape turbinate to slightly pyriform; size usually large, 13-18 cm. ht., 12-16 cm. diam., weight 1-3½ lb., D/H index 0.89-0.96 (av. 0.91); colour light greenish yellow; surface smooth, papillate like common shaddock, pitted; base very slightly depressed, very low-collared, ribbed or rugose, apex almost rounded to slightly truncate; areole absent, stylar scar present, even rarely, depressed in a small pit, style not persistent.

*Internal fruit characters.* Rind thick, thickness 2-3 cm., adherence somewhat strong; oil glands conspicuous, number 12-18 per ¼ sq. cm., primary glands spherical and pyriform, size usually 2-4 mm. long, 1-3 mm. broad, glandular layer about 5 mm. thick, aroma distincly of shaddock; mesocarp (albedo) white, spongy, bitter; axis hollow or semi-hollow, diameter at the centre 2.2-2.8 cm.; segments 9-13, adherence medium, septa somewhat coriaceous, tough usually not broken at the core, dorsal contour obtuse or truncate; pulp-vesicles white, texture coarse,arising only from the dorsal side, shape usually fusiform, melting; juice abundant, 4-7 oz., very sour, white, not bitter; seeds many, 20-60 per fruit, shape usually cuneate one side smooth, other side rough with anastomosing lines, rounded end rough, micropylar end fiat, size 1.8-1.9 cm. long, 1-1.2 cm. broad, 6-7 mm. thick, testa cream white, tegmen light brown, chalazal spot brown, cotyledon white.

This is a probable hybrid apparently between *C aurantium* and *C. grandis.* The shape and the aroma of the crushed leaves are distincly like those of the Seville orange (*C. aurantium* Linn.) while pubescent branches, large size of the fruit and the flowers are indicative of association with the *pummelo* (*C. grandis* Osbeck). The fruits are useless for any commercial purpose.

# Appendix

Chemical analysis of the Citrus fruit varieties classified in the text.

| S. No. | Name of species | Local name of the variety | Citric acid per 100 c.c. of juice | Grape sugar per 100 c.c. of juice | Cane sugar per 100 c.c. of juice |
|---|---|---|---|---|---|
| 1. | *C. medica* Linn. | Bira-jora | gm. Not analyzed | gm. | gm. |
| 2. | Do | Soh-manong | Do | – | – |
| 3. | Do | Mitha-jora | Do | – | – |
| 4. | *C. limon Burm.* | Assam-lemon | 7.869 | 0.127 | 0.382 |
| 5. | Do | Pati-lebu | 6.376 | 0.253 | 0.955 |
| 6. | Do | Jora-lenga | 7.869 | 0.572 | 0.317 |
| 7. | Do | Godha-patilebu | Not analyzed | – | – |
| 8. | Do | Kata-jamuri | 3.934 | 1.336 | 0.254 |
| 9. | Do | Elachi-lebu | Not analyzed | – | – |
| 10. | Do | Naya-changney | Do | – | – |
| 11. | Do | Soh-long | 5.695 | 1.718 | 0.254 |
| 12. | Do | Soh-synteng | 5.088 | 2.418 | 0.252 |
| 13. | Do | Pani-jamir | 1.221 | 4.709 | 3.309 |
| 14. | *C. jambhiri* Lush | Soh-myndong | 4.273 | 4.963 | 2.157 |
| 15. | Do | Soh-jhahlia | 7.191 | 2.927 | 1.145 |
| 16. | Do | Kata-jamir | 6.444 | 2.354 | 0.763 |
| 17. | Do | Sinduri-nemu-tenga | 5.869 | 2.418 | 1.782 |

*Contd...*

**Appendix** ***Contd...***

| S. No. | Name of species | Local name of the variety | Citric acid per 100 c.c. of juice | Grape sugar per 100 c.c. of juice | Cane sugar per 100 c.c. of juice |
|---|---|---|---|---|---|
| 18. | Do | Nemu-tenga | 4.545 | 5.790 | 1.272 |
| 19. | Do | Mitha-tulia | 1.153 | 4.836 | 1.909 |
| 20. | *C. karna* Raf. | Soh-sarkar | 6.105 | 2.418 | 0.636 |
| 21. | *C. aurantifolia* Swingle | Abhayapuri lime | 6.784 | 0.763 | 0.382 |
| 22. | Do | Karimganj-lime | 8.819 | 0.127 | 0.891 |
| 23. | *C. limetta* Risso | *Mitha-kaghzi* (Sweet lime) | 0.203 | 9.209 | 1.736 |
| 24. | *C. reticulata* Blanco | Soh-niamtra | 0.339 | 4.581 | 6.618 |
| 25. | Do | Soh-umkdait | 0.746 | 3.370 | 3.620 |
| 26. | *C. reticulata* Blanco | Naga-santra | gm. Not analyzed | gm. | gm. |
| 27. | Do | Soh-siem | 1.967 | 4.380 | 3.020 |
| 28. | Do | Kapura-tenga | 1.763 | 2.736 | 9.609 |
| 29. | Do | Narangi | 7.123 | 1.527 | 3.736 |
| 30. | *C. nobilis* Lour | Jeneru-tenga | 1.289 | 2.770 | 2.420 |
| 31. | *C. indica* Tanaka | Indian wild orange | Not analyzed | – | – |
| 32. | *C. sinensis* Osbeck | Soh0niangriang | 2.374 | 9.654 | 4.006 |
| 33. | *C. aurantium* Linn | Karun-jamir | 4.545 | 4.645 | 1.781 |
| 34. | Do | Gondh-huntra | 2.374 | 5.472 | 1.617 |
| 35. | *C. grandis* Osbeck | Dowadi (pummelo) | Not analyzed | – | – |
| 36. | Do | Nowgong (pummelo) | 1.221 | 5.027 | 3.499 |
| 37. | Do | Jorhat (pummelo) | 1.492 | 5.090 | 3.563 |
| 38. | Do | Khanapara (pummelo) | 0.814 | 3.627 | 2.863 |
| 39. | Do | Kamrup (pummelo) | 0.746 | 3.372 | 1.526 |

***Contd...***

**Appendix *Contd...***

| S. No. | Name of species | Local name of the variety | Citric acid per 100 c.c. of juice | Grape sugar per 100 c.c. of juice | Cane sugar per 100 c.c. of juice |
|---|---|---|---|---|---|
| 40. | Do | Khasi (pummelo) | 0.814 | 3.818 | 2.609 |
| 41. | Do | Burni (pummelo) | Not analyzed | – | – |
| 42. | Do | Zemabawk (pummelo) | 1.153 | 7.063 | 2.990 |
| 43. | Do | Aijal (pummelo) | 2.374 | 6.745 | 4.964 |
| 44. | *C. megaloxycarpa* Lush | Bor-tenga | 2.849 | 3.181 | 4.199 |
| 45. | Do | Hukma-tenga | 4.002 | 1.463 | 3.935 |
| 46. | Do | Holong-tenga | 3.459 | 0.381 | 2.418 |
| 47. | Do | Jamir-tenga | 4.273 | 3.945 | 0.700 |
| 48. | *C. ichangensis* Swingle | Ketsa-shupfu | Not analyzed | – | – |
| 49. | *C. latipes* Tanaka | Soh-kymphorshrieh | 5.630 | 2.354 | 2.291 |
| 50. | Do | Soh-shyrkhoit | Not analyzed | – | – |
| 51. | *C. macroptera* Mont | Sat-kara | 7.081 | 0.381 | 0.254 |
| 52. | Do | Tith-kara | Not analyzed | – | – |
| 53. | *C. assamensis* Dutta & Bhatta, Sp. Nov. | Ada-jamir | 6.919 | 1.909 | 0.441 |
| 54. | Probable hybrids | Soh-khyllah (Tangelo type, hybrid No. 1). | 1.628 | 6.427 | 1.909 |
| 55. | Do | Soh-khyllah (At-ani of Bonavia, hybrid No. 2). | 1.224 | 4.050 | 3.420 |
| 56. | Do | Soh-khyllah (Tample orange type, hybrid No. 3). | 1.289 | 4.080 | 4.560 |

*Contd...*

**Appendix** ***Contd...***

| S. No. | Name of species | Local name of the variety | Citric acid per 100 c.c. of juice | Grape sugar per 100 c.c. of juice | Cane sugar per 100 c.c. of juice |
|---|---|---|---|---|---|
| 57. | Do | Hash-khuli (hybrid No. 4) | Not analyzed | – | – |
| 58. | Do | Sarbati (hybrid No. 5) | 3.459 | 4.263 | 1.718 |
| 59. | Do | Nichols-Roy (hybrid No. 6) | 2.850 | 1.840 | 4.170 |
| 60. | Do | Dewa-tenga (hybrid No. 7) | Not analyzed | – | – |

# Index

**T**



**V**



**Z**